Muhammed Abed Mazeel

Formation Damage. Experimental Work. Laboratory Identification and calculation of Formation damage

disserta
Verlag

Mazeel, Muhammed Abed: Formation Damage. Experimental Work. Laboratory Identification and calculation of Formation damage, Hamburg, disserta Verlag, 2024

Buch-ISBN: 978-3-95935-630-5
PDF-eBook-ISBN: 978-3-95935-631-2
Druck/Herstellung: disserta Verlag, Hamburg, 2024
Cover: © disserta Verlag

Bibliografische Information der Deutschen Nationalbibliothek:
Die Deutsche Nationalbibliothek verzeichnet diese Publikation in der Deutschen Nationalbibliografie; detaillierte bibliografische Daten sind im Internet über http://dnb.d-nb.de abrufbar.

Hermannstal 119k, 22119 Hamburg
http://www.disserta-verlag.de, Hamburg 2024
Printed in Germany

Table of Contents

List of Figures

List of Tables

ACKNOWLEDGEMENTS

I would like to thank the Institute of Petroleum Engineering, University of Clausthal-Zellefield in Germany who gave me the admission to public this book, which is big part of it from my PhD thesis.

The books have been published are five books in different subjects and this is the sixth book. Two PhD thesis and one master thesis. The revenue from those books donated to the poor and cancer sick children.

Dr.-Eng. Dr. Management Muhammed Abed Mazeel

1. INTRODUCTION

The permeability and porosity of the rock formations are of great importance to the reserves and productivity of oil or natural gas reservoirs. Porosity, low water saturation and permeability are key parameters to move out the fluid of the pore space into the wellbore for production. Obviously, pressure in the reservoir is important to be able to produce the fluids. Additionally, mineralogy and susceptibility of the formation are also important.

During drilling and the subsequent treatment, the original permeability of the formation near the well and the respective distant area can be reduced through the interaction of the drilling mud and the treatment fluids. This reduction of the permeability of the reservoir is referred to as “formation damage”.

In order to stabilise the drill hole and prevent uncontrolled influx from the formation, the density of the mud will be adjusted so that the hydrostatic pressure of the mud pile is greater than that of the formation. Through the resulting positive pressure difference, solid substances or filtrates of the drill mud and the treatment fluids being used can penetrate into the pores of the rock formations; this can lead to physical and/or chemical reactions with the substance contents of the pores and the rock-formations. In a hydraulic fracturing treatment, a fluid is injected with such high pressure into the well that it results in cracks developing in the formation. The permeability of the reservoir is thus improved through artificially produced, highly permeable flow canals. This is of value as long as the supporting material introduced is not destroyed through the overlapping pressure of the rocks and is fully embedded in the crevices.

Underground or surface installations can be damaged through the dragging along of sand in the process of petroleum and natural gas exploitation as well as after treating operation of the reservoir. The mechanical retention of sand in the well takes place through the sand filter (gravel pack); the inflow of gravel material takes place with the help of gravel formation fluids, with polymers mixed into them. The formation can solidify through the sand filter, which leads to a binding of the sand from the reservoirs. The term borehole acidulation is understood as the chemical removal of the wall coating of the well (filter cakes) and the expansion of the flow routes in the formation. Diluted hydrochloric acid (among other acids) can be put in, in order to dissolve the lime reservoirs; in the process of fracturing, the acid can additionally expand the flow route.

The area of the formation that is damaged through the interaction the drill mud and the treatment fluids can amount to between a few centimetres and many metres. With this, the depth of the perforation is normally not enough to surmount the damaged zone. To avoid expensive well treatment aimed at increasing the permeability of the formation, fluids appropriate to the formation must be put in earlier at the drilling and treatment phases. Hence, the adjustment of these fluids to the corresponding conditions of the reservoir is the aim.

Since according to /1/ the API filtration values of the drill mud and the treatment fluids do not indicate their filtration properties under well conditions, this book is extended experimental program to Author PhD thesis and will examine drill mud and treatment fluids under simulated well conditions with regard to their influence on different permeable rock samples.

2. BASIC RESERVOIR THEORY

Porosity is the measure of the void space in a rock and usually expressed as a percent. The amount of intergranular porosity will depend upon uniformity of particles and the packing arrangement of the particles. Porosity can be classified as effective porosity (continuous or interconnected porosity), non-effective porosity (discontinuous or isolated porosity) and total porosity (the sum of all the porosity). It can further be classified as primary or secondary porosity. Primary porosity is that porosity formed when sand grains or carbonates are deposited, compacted, and cemented over geological time. Secondary porosity is that porosity formed after deposition, compaction, etc by ground water moving through the rock dissolving some of it and by the earth stresses causing movement and cracking or natural fracturing of the rock, all of which provides void spaces, i.e. porosity. In general porosity in sandstone is normally primary porosity and much of porosity in limestone can be secondary. There are two principal reasons for this. One is that limestone and dolomites are usually more “brittle” (les elastic) than sandstones, i.e., they are not as elastic as sandstones and will fracture rather than give as easily as sandstone. The other is limestones and dolomites are more chemically reactive than sandstones. As ground water moves through these carbonates it will dissolve and possibly reprecipitate carbonate. When the dissolution rate exceeds the precipitation rate, porosity is formed. Vugular porosity in the carbonates is evidence of this type of secondary porosity development. More extreme evidence of this is the caves and the caverns found in many limestone formations. Porosity gives a formation reservoir capacity of importance the oil and/or gas saturation (So/Sg) in a reservoir which is the percentage (or fraction) of the pore space filled with hydrocarbon. It is more common to refer to the water saturation, Sw, as that portion of the pore space filled with formation (connate) water. Most all sedimentary rock is in contact with water at the time of deposition or at least before coming in contact with oil or gas. Therefore, most reservoirs begin with 100% Sw and are water wet. As hydrocarbons form and begin to migrate they can displace some of the water and be trapped in reservoirs. The final hydrocarbon and water saturation will depend on several factors. One of these factors would be the size of the pores that from which the water is being displaced. The smaller the pore, the higher the capillary pressure or affinity the water has for the pore. The higher this pressure the more difficult it would be for hydrocarbons to displace the water. Porosity can be determined from well logs and analysis of formation samples. Water saturation can also be determined by logs and core analysis.

Permeability is a measure of the ease which fluid can move through interconnected pore spaces of the rock. The rock is more or less permeable depending upon whether the rock will let fluid pass through it with greater or lesser ease. This can be related to pipelines where larger, straighter lines let fluid flow through more easily than smaller lines with more turns. Many rocks are impervious to movement of water, oil, or gas, even though they may actually be quite porous. Examples of these are shale's, some chalks and some highly cemented sandstone. Darcy law applies to any fluid flowing in any direction through a porous material as long as the fluids are Newtonian and in laminar flow. It is expressed in equation form as follows for radial flow of an incompressible fluid (eq. 1 and 2):

$$K = \frac{Q . \mu . \ln\left(\frac{re}{rw}\right)}{7.707\ h . (Pe - Pw)} \qquad \text{Eq. 1}$$

or

$$\mathbf{Q} = \frac{\mathbf{7.707\ k.h.(Pe - Pw)}}{\mathbf{\mu . \ln\left(\frac{re}{rw}\right)}} \qquad \text{Eq. 2}$$

Where: k= permeability, darcies

Q= Flow rate, barrels per day

μ= Flowing fluid viscosity, centipoises

re= Drainage radius for the well, feet

r_w= Wellbore radius, feet

h = Thickness of the zone contributing to flow, feet

P_e= Reservoir pressure, psi

P_w = Pressure at the wellbore, psi

Reservoir rock having an average permeability as great as 1 Darcy is found only in small portion of the reservoirs containing petroleum deposits. The usual measure of rock permeability, therefore, is in millidarcies (md) or thousandths of Darcy. The above equation assumes that only one fluid is in the reservoir and it is the fluid flowing, in laminar flow. As we know, we will usually have at least two fluids and perhaps three, oil, gas, and water in many of the reservoirs. For this reason, we should utilize a relative permeability to define the flow of more than one fluid through a rock (Figure 1).

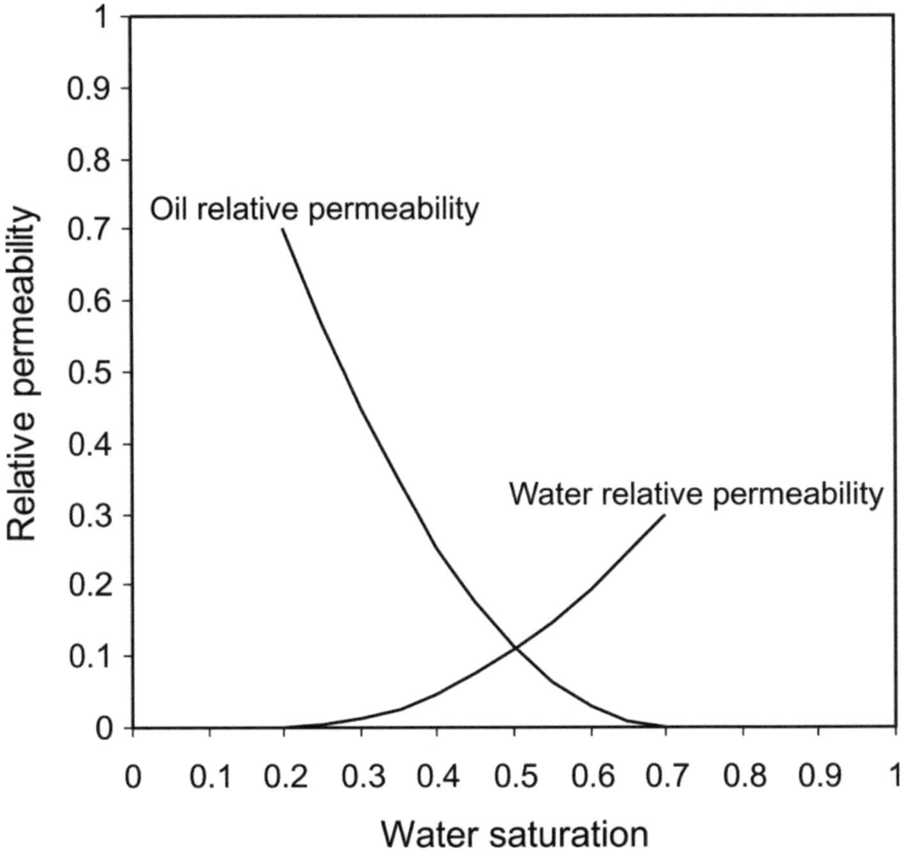

Figure 1: Oil-Water Relative Permeability

There are several factors that influence porosity and permeability. The porosity and permeability of sandstone depend upon size and shape of grains, variations in size of the grains, arrangement in which grains were laid down and compacted and the amount of clay and other materials which cement the sand grains together. Size of the sand grains with make up a rock do not influences the amount of porosity occurring in the rocks. However, variation in sand grain sizes considerably influence the amount of porosity Figure 2. The porosity between large spheres is the same as that between small spheres, provided that the spheres are arranged in the same manner. If they are arranged in cubic order, porosity is about 48 percent, whereas, if they are arranged in rhombohedral order, porosity is about 26 percent. Where spheres (grains) of varying sizes are packed together porosity will be significantly reduced. Since there is some correlation between porosity and permeability, by porosity being reduced, permeability will probably also reduced. If sand grains are elongated or flat and packed with their flat surfaces together, porosity and permeability may both be low. Permeability along the flat surfaces will be higher, however, than in the direction that is perpendicular to, or across, the flat surfaces of the grain. In reservoir, the horizontal permeability is usually higher than the vertical permeability (Figure 2A,C).

Figure 2 A and B

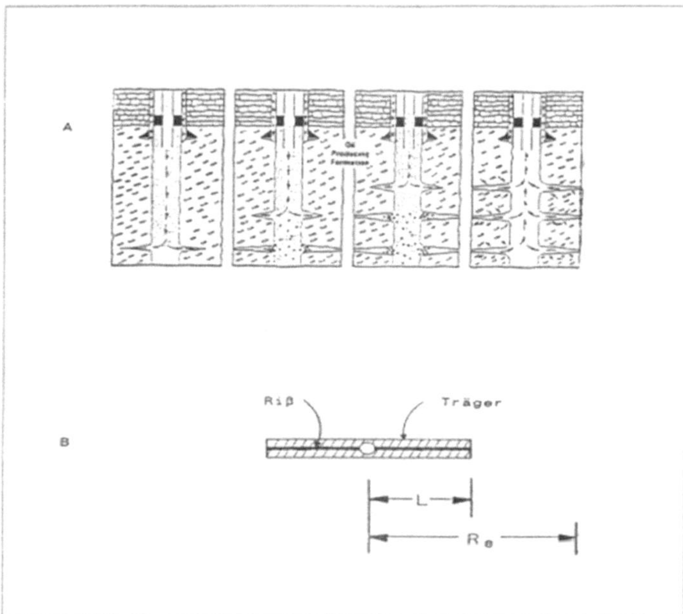

Figure 2C

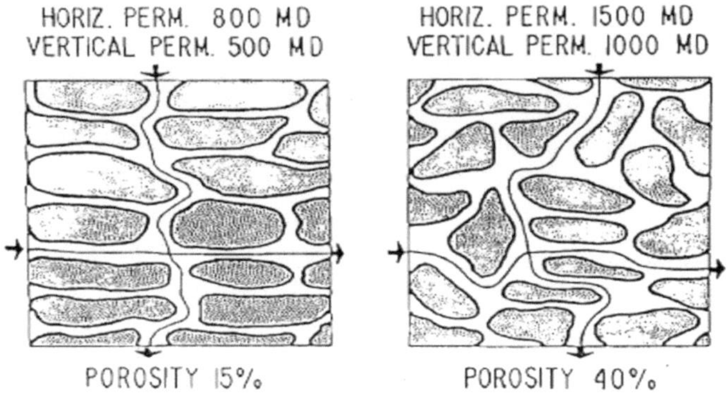

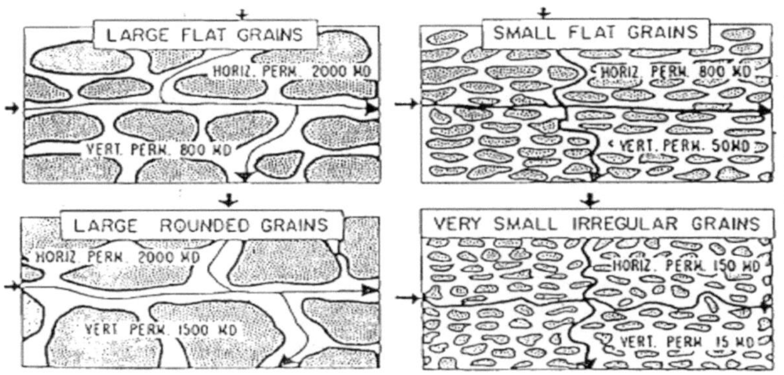

Figure: 2 A, B, C: Effects of Shape and Size of Sand Grains

This is caused by the process of sedimentation, where the sand grains are washed to their final position of the flattest sides in horizontal position, due to thin shale laminations and due to compaction forces from the overburden, compacting the grains and cementing materials more in the vertical direction than horizontal. Obviously, the amount and location of cementing material will have an effect on the porosity and permeability.

Limestone's (except for granular oolitic types which may be similar in character to sandstone) are composed of more densely spaced particles and porosity is found between microscopic crystals in the rock, solution void spaces caused by the movement and solvent action of ground water and other spaces such as natural fractures caused by tectonics /87/.

Porosity and permeability however have the same meaning in limestone as in sandstone, although they are usually extremely irregular and much more difficult to measure and analyse in limestones. Porosity development in dolomites can vary, but as limestone is converted to dolomite through dolomitization, porosity can develop.

When dealing with the flow of more than one fluid through a rock, relative permeability, the ability of one fluid or all fluids to flow relative to one fluid flowing, must be considered. If we consider an oil reservoir at high pressure, where gas has not been allowed to come out of the solution and the water saturation is low, only oil will flow. Following production, reservoir pressure will reduce (unless there is a strong water drive), gas can evolve from the oil and being to fill a portion of pore space. If the pressure decline is slight the gas that forms will not be sufficient to form continuous pore to pore gas saturation, still only oil can flow from the pore spaces. However, just by gas being in the pore space it can slow the oil flow rate. If reservoir pressure is further reduced the small bubbles of gas in each pore will grow in size until a continuous gas phase is established along the pore network and the production will be oil and gas. As pressure continues to decline and gas saturation increases, more and more gas and less oil is produced. Thus, the relative permeability has shifted to a higher relative permeability to gas as reservoir pressure was reduced.

Relative permeability is defined as the rate of flow of one fluid, such as oil, through pore channels in the presence of a second fluid, such as gas, as compared to the rate of flow under the same pressure drop conditions, if only one fluid were present. Figure 1 shows an example relative permeability curve. In this example the absolute permeability, i.e., the permeability of the rock tested with only air in the pores and the air flowing was 214 md. The curves are for oil and water relative permeability's at various water saturation (the wetting phase) and no gas in the rock. With water saturation of 35% the permeability to oil is approximately 100 md (0.46 Kro *214 md). As the water saturation increases the relative permeability to oil goes down and oil permeability is reduced. At the 35% water saturation there would be some permeability to water (0.03Krw * 214 md), approximately 6-7 md and some, although small, amount of water produced.

In addition to having reserves, i.e., porosity with hydrocarbon saturation and permeability there must be sufficient energy in the reservoir to bring the hydrocarbon to the wellbore. That energy is reservoir pressure. At the time of deposition or shortly thereafter, the formation was in contact with water and was buried with that water or other water which be equal to the hydrostatic pressure of a column of water, usually sea water, to the depth of the reservoir. Sea water, approximately 3.5% salt, has a gradient of approximately 0.44 psi/ft, 9.95 kPa/m at a depth of 5000 feet, 1524 m, a formation would have a reservoir pressure of 2200 psi, 15,164 kPa, if normally pressured. This is the pressure at which oil and gas migrated into the reservoir and would be the energy for production. Of course, this situation could be changed by geological time and the reservoir pressures could be different than the normal gradient. For example, some reservoirs are found with virgin reservoir pressures of 0.3 psi/ft, 6.6 kPa/m, to some that are greater than 0.8 psi/ft, 17.6 kPa/m. The lower gradient may be due to some of the hydrocarbon escaping over geological time and the higher possibly due to shallower burial (erosion of overburden) since the reservoir was invaded by the hydrocarbon and the "trap" was sealing, preventing the hydrocarbon from escaping. These higher than normally pressured reservoirs are said to be geopressured. When reservoirs are geopressured, the pore pressure is supporting a portion of the overburden load and as pore pressure is reduced the matrix of the rock must support this additional load. Because of this the reservoir pore structure may change resulting in reduced permeability as pore pressure is reduced. This reduced permeability can be a skin damaged initially seen in the near wellbore area. This is a skin damage that can be caused simply by producing the well. Without knowledge of the mechanical properties of the formation and the stress environment, it would be easy to mistake this type of “damage” for other type of damage and the wrong treatment selected. This stress sensitivity seen in geopressured formations can also occur in normally pressured formations where in-situ stresses are high. The stress working on and around the wellbore in horizontal wells can be very high causing a reduction in deliverability as the pore pressure is reduced. Because of the importance of understanding this form of “damage” it is important to take a basic look at rock mechanics and how formation behaves under pressure. An absolute understanding of how they behave becomes a very detailed science, however, there are some basic we can review which will give a good understanding of rock mechanics. Rock mechanics properties vary considerably depending upon the makeup of the rock and the forces to which the rock is subjected. Obviously, shale behaves differently than a dense dolomite; unconsolidated sandstone behaves differently than well consolidated sandstone, etc. Rock mechanical properties can measure in the laboratory on core samples or more indirectly with logs in the wellbore. Two rock mechanical properties important to a basic understanding of rock behaviour under stress are Young´s Modulus of Elasticity and Poisson´s ratio. Some general ranges for Young´s Modulus of Elasticity for the common formation types we work with are as follows:

Table 1: Ranges for Young´s Modulus of Elasticity

Shale approx.:	1-3,000,000psi or 6,900-20,670 MPa
Sandstone approx.:	5,000,000 psi or 3,445 MPa
Limestone/Dolomite approx.:	8-10,000,000psi or 55,000-68,900 MPa

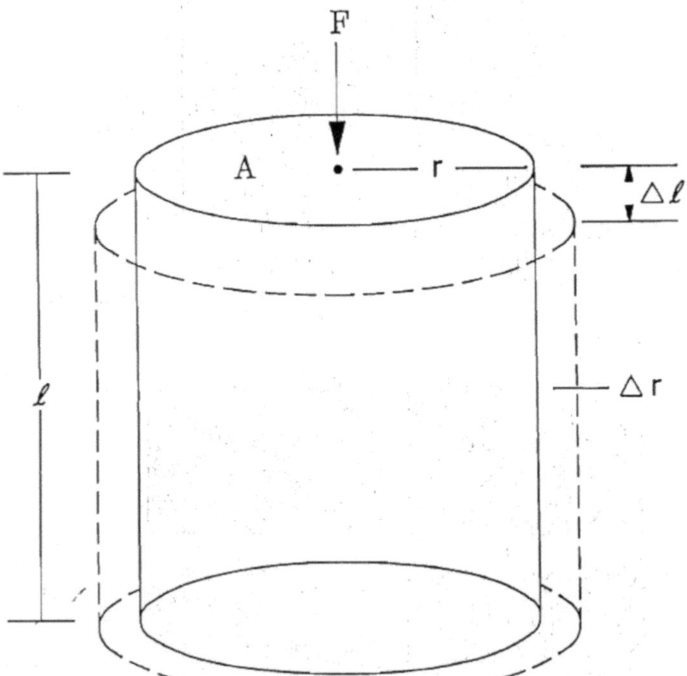

Figure 3: Core sample Loaded Axially Showing Strain, Resulting from Stress, F/A

If we think of a core sample being loaded axially, parallel to the length of the core we would expect the core to deform under the load as shown in Figure 3. This relationship between stress and strain helps to define rock´s behaviour when it is under load.

The mount of deformation for a given load helps us define the elastic properties of the rock. The rock initially deforms elastically as the load is applied. If the load is released, we would expect the core to rebound or return to its original shape if it is perfectly elastic in behaviour. The load of force being applied, F, is over some surface area, A. We define this force per unit area as stress. Or, stress=force per unit area, δ=F/A, usually expresses as pounds per square inch, psi, in English units and Pascal, pa, in metric units. Stress can be referred to as total stress or effective stress. The effective stress, δ´, is equal to the total stress, δ less some portion of the formation pressure, Po, as follows (eq. 3):

$$\delta' = \delta - \alpha \, Po \quad \text{Eq. 1}$$

Effective stress is defined as the stress carried by the rock grains. Since formations are made up of grains or a matrix with pore spaces that contain fluid, the fluid can carry a portion of the stress. The relation between pore pressure and stress is defined by a poroelastic constant (Biot´s constant), α it basically describes the efficiency of the fluid pressure in counteracting the total stress. Its value is dependent upon the pore geometry and the physical properties of the constituents of the fluid and solids that make up the rock. It is better described by its relation to the compressibility of the rock grains and the total rock compressibility (eq. 4):

$$\alpha = 1 - \frac{Cg}{Cb} \quad \text{Eq. 4}$$

Where: Cg= compressibility of the grains

Cb= bulk rock compressibility (dry)

The larger the value of α the easier it is to compress the rock. In the case of a very low porosity rock, α can be substantially less than 1, implying that pore pressure has less effect on the stress carried by the formation.

3. ROCK MECHANICS

Rock mechanics is the study of how rocks behave under pressure. An absolute understanding of how they behave becomes a very detailed science; however, there are some basics we can review which will give a good understanding of the fracturing process, fracture behaviour and fracture geometry as well as resulting fracture flow capacity. Rock mechanical properties vary considerably depending upon the makeup of the rock and the forces to which the rock is subjected. Obviously, shale behaves differently than a dense dolomite; unconsolidated sandstone behaves differently than well consolidated sandstone, etc. Rock mechanical properties can be measured in the laboratory on core samples or more indirectly with logs in the wellbore. The rock mechanical properties important to a basic understanding of the fracturing process are Young's Modulus of Elasticity and Poisson's ratio. If we think of a core sample being loaded axially, parallel to the length of the core we would expect the core to deform under the load as was shown in Figure 3.

From the loading of a rock sample the resulting change in shape, (shorting of core, temporary deformation) is the strain. Strain can be thought of as the change in length of the core, temporary deformation) is the strain. Strain can be thought of as the change in length of the core as compared to the original length as shown in Figure 3. Also shown on this figure is the deformation of the core perpendicular to the direction of the stress being applied. This lateral deformation can be compared to the axial deformation of the core and the Poisson´s ratio defined. Poisson´s is defined as the lateral strain/axial strain or the deformation perpendicular to the core as compared to the deformation parallel to the core.

Poisson´s ratio may vary from approximate 0.1 to 0.35 depending upon the type of formation and its properties, clay content, etc.

The formations with which we deal are subjected to three principle stress. One is in the vertical direction and two in the horizontal direction, as shown in Figure 3. The two horizontal stresses are rarely equal and differences in the two have been reported as high as 800 psi, 5,500 KPa. The anticipated vertical stress, δv, is approximately equal to 1-1.1 psi/ft, 22.6-24.9 KPa/m. Thus, for a 10,000 ft, 3048.8 m, well the stress acting in the vertical direction would be 10,000-11,000 psi, 68.9-75.9 MPa. In a totally tectonically relaxed area, i.e., the only stress being applied on the formation is that from above, then the stresses in the horizontal direction is that which is transfessed via Poisson´s ratio (this is rarely the case). The least principal stress, δh, can be measured by breaking the formation down and measuring the fracture closer pressure. This is the least principle stress. This is done frequently in relation with hydraulic fracturing treatments. The least principle stress measured in this way has been seen to be in the range of 0.6 to 0.9 psi/ft, 13.56 to 20.34 KPa/m. Values outside this range have been measured, and however, it is less common. Where the intermediate stress, δh, has been determined it has been found to be as high as 800 psi, 56 Kg/cm2, greater than the least principle stress. This magnitude was reported for the Ekofisk field in the North Sea. As an example, a 10,000 ft, 3048.8 m formation would have a vertical stress of 10,000 psi, 68.9 MPa and a least principle horizontal stress of perhaps 7000 psi, 48.2 MPa. The intermediate stress would be expected to be greater than the least principle stress, perhaps as much as seen in Ekofisk. When we drill a vertical well into a formation with a stress environment as discussed here, the stress acting on the wellbore are from the horizontal stresses and unless we are drilling in highly tectonically active area we don´t expect these stresses to give us wellbore stability or other problems. However, when we drill deviated holes with angels as high as 90° (horizontal wells) the effect of the vertical stress is felt on the wellbore and near wellbore area. Because of the difference between the two stresses acting on the wellbore we would expect to have a wellbore deformation effect as shown in Figures. With a vertical well δy would be the intermediate stress, δh and δx, would be the least principle stress, δh. For a horizontal well δy would be the vertical stress, δv, and δx, would be the intermediate stress, δh for a horizontal well drilled in the direction of the least principle stress, δh, which is normally the case. For a vertical well with a small difference between the two horizontal stresses we do not expect much additional stress around the wellbore. However, as stated above, with a horizontal well the difference between the stresses working on the wellbore can be much greater causing much higher stress around the wellbore. As Figure 3-7 indicates, where the maximum stress is working on the wellbore (points "a") a tensile force is created and at points "b" a compressive force is created. The greater the difference between the two stresses working on the wellbore the greater the magnitude of the tensile and compressive forces. Many horizontal wells are drilled in the direction of the least principle stress to connect as many nature fractures as possible this is also the horizontal direction of greatest wellbore stability since the difference between the two stresses acting on the wellbore is less than if the wellbore was drilled in the other direction, towards the intermediate principle stress.

Some points need to be made here with this brief discussion of rock mechanics, stresses and wellbore stability. One is that in horizontal wells the wellbore stability issue may require that a high mud weight, much higher than the reservoir pressure, be used to prevent wellbore collapse while drilling. Figure 3-7 shows an example of wellbore stability curve. When stability curves as the figure shows, as the wellbore is deviated from the vertical to horizontal orientation (0 to 90°) the stress on the wellbore increase and the window of acceptable mud weight gets narrower. If this overbalanced drilling condition requirement is necessary, it can cause a lot of fluid loss and formation damage. Another point is that these stresses working in the near wellbore area can cause pore collapse and natural fracture closure resulting in reduced productivity, damage as discussed earlier. Also, as discussed, the effective stress, δ´, is equal to the total stress, less some portion of the reservoir pressure, Po, as follows (eq. 5):

$$\delta' = \delta - \alpha Po \quad \text{Eq. 2}$$

Where α the poroelastic constant, is high, approaching 1, the effect of reducing pore pressure significantly increases the effective stress and can cause pore collapse and natural fracture closure.

4. FRACTURE MECHANICS

In most areas the vertical stress working on a formation is equal to the weight of the overburden or about 10 -11 psi/ft, 226 - 249 KPa.m of depth. In a totally relaxed tectonic area, i e , the only stress being applied on the formation is that from above, then the stress in the horizontal direction is that which is transferred via Poisson's ratio. In all areas there are three principle stresses, one in the vertical direction and two in the horizontal direction as was shown in Figure 3. In most all cases the two horizontal stresses are not equal. This implies that there are forces at work other than just the stress coming from the weight of the overburden. This is the usual case Stresses related to tectonic activity; monntrun building, faulting, etc cause these stresses to be dissimilar.

However, we still expect three principal stresses as shown in Figure 3. When we drill a well in the formation the wellbore is deuced by these stresses and the stresses around the wellbore will cause some degree of hole deformation. This was shown in Figure 3-7, exaggerated to point out these effects. This deformation will be due to the difference in tectonic activity, etc. If the pay zone stress is significantly less than the stress in the zones above and below then the fracture will preferentially stay in the pay zone as long as the pressure (in the fracture) during the fracturing treatment does not exceed the pressure difference between the pay and the other zones. So, we see that pressure in the fracture is an important parameter in the final outcome of the fracturing treatment and something that should be monitored closely during the treatment. Another factor influencing fracture geometry and pressure in the fracture is fluid leakoff. As the fluid leaks off into the formation the fluid volume creating fracture growth lessens and effects pressure in the fracture. This is affected in two ways, one is the smaller frac fluid volume generating a lower pressure and less fracture growth, the other is the fluid leaking off into the formation pores creates a pseudo pore pressure which we must work against. Other factors are the viscosity of the fracturing fluid which may include propping agent, the type of fluid in the formation pore space and the injection rate. Obviously, the formation permeability and porosity would play a role in the amount of fluid leaking off. So, the geometry of the fracture is influenced by many factors. To better design hydraulic fracturing treatments, the more of this type of information available the more coddent one can be in planning a profitable treatment. The geometry of concern is the height of the fracture, the width of the fracture and the length of the fracture. The width of the fracture is also related to the pressure in the fracture and is a function of the injection rate, the viscosity of the fracturing fluid, length of the fracture and the elastic properties of the formation. The basic information necessary to get a good idea of how a fracture may grow (will the fracture stay in zone?), is the stress variation through and on both sides of the pay zone. Stress measurements are made in several ways. The best approach is to conduct a microfrac treatment in the zones of interest. A microfrac consists of fracturing the zone of interest with a small volume of frac fluid at a low rate, shutting down and measuring versus time relation breaks over indicating fracture opening. This fracture opening pressure will be an approximation of fracture closure pressure. The difference will be due to possible friction through the perforations and the closed fracture and to back stresses developed from fluid leakoff. However, if the same technique and analysis method is used for several intervals, the difference in stress between them should be fairly accurate. In all cases, it is preferred to obtain downhole pressure data. Also, downhole shut-in is preferred in the no flowback case to minimize wellbore storage effects which can mask the required early time pressure falloff response. What we have commonly called the instantaneous shut-in pressure (ISIP), which is the instantaneous downhole pressure when the rate goes to zero, is called the closure stress. This is, in fact, an upper boundary for the closure stress, although it may be very close in some cases. Following the moment of ISIP, there will be a period of linear flow as the fluid in the fracture leaks off into the foundation. The extent of this period and the magnitude of the pressure change will depend on the formation properties and the rate and pressure history of the microfrac up to that point. For a very low permeability rock, the leakoff time might be very long, but the pressure changes significantly, so ISIP is a valid approximation. The falloff after ISIP may even be dominated by communication behind pipe. The change is relative to the shut-in time. The pressure will also change linearly on a plot of pressure versus the square root of time, but with an arbitrary slope. The linear flow region can be identified on the log-log plot and a straight line can be fit to the corresponding time interval on the square root time plot. The point at which the data breaks away from this line corresponds to the closure stress replace frac sand at pressures greater than about 5000 Psi, 345 MPa. The glass beads were the only high strength proppant available for several years and they were widely used. The problems with glass beads were that they were not that much stronger than sand, they tended to break easier if they were scratched, in salt water environments, etc. Additionally, when they failed basically went to powder resulting in very low flow capacity. When sand fails it tends to break into several pieces rather than shatter into a powder and because of this sand result in higher flow capacity at high closure pressures than glass beads. The first true high strength propping agent developed was sintered bauxite in the mid 1970's. This proppant proved to be extremely strong, stronger than most formations and is still available today. It had a density much greater than sand (37 compared to sand at 265 &cc) and because of this it was more difficult to transport in the fracturing fluids. The fracturing fluids must be designed to compensate for this. Since the introduction of sintered bauxite several propping agents have been developed which give improved performance and economics for use in the intermediate closure pressure range. The common materials in use today for propping agents are sand, resin coated sand, light weight and intermediate strength proppants and sintered bauxite.

5. FORMATION DAMAGE

The term Formation Damage is understood as the reduction of the permeability of a petroleum or natural gas reservoir with the accompanying reduction of its productivity.

Formation protection comprises all technical measures with mud that are helpful to the conservation and restoration of the original permeability of the formation rock /10/.

In the process of getting through the depth of a formation, the drill mud penetrates into the formation under the differential-pressure ΔP (the difference between the mud pile pressure and the pore contents pressure) over a certain period (the so-called mud spurt loss). In doing so, the solid substances of the drill mud become entangled in the narrow spots of the flooded pore channels, and an internal filter-cake develops Figure 4.

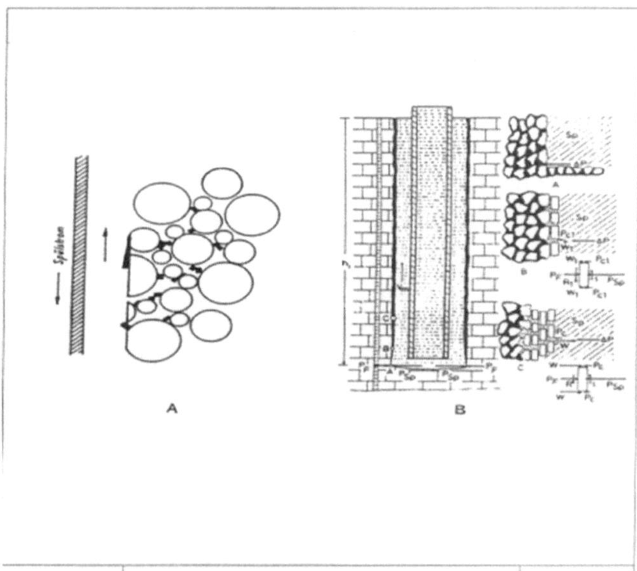

Figure 4: Internal and External Filter Cake

Figure 5 shows the result of a series of /3/ analysis that were carried out with regard to the mud spurt invasion into a highly-permeable limestone. A series of recordings based on the return radiation from the nucleus components around neutron logs, presents the difference phases of the penetration of fluids in 2-second intervals. Different tone concentrations are characterised with different colour tones. Blue represents the highest tone concentration, while the decrease in tone concentration is recorded from reddish-brown to white.

The raster records show that the solid substances penetrate into the highly-permeable formation within seconds. However, after 8 seconds, a it is hardly possible to establish any further change.

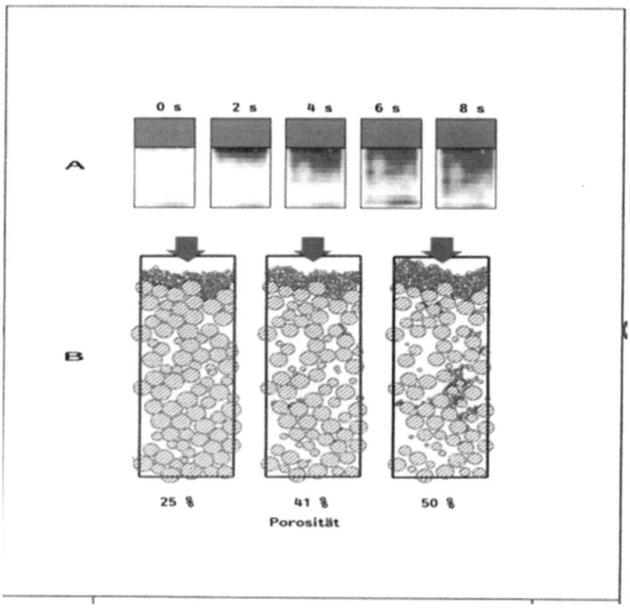

Figure 5: Raster Records for Different Porosities Formation

Figure 6 illustrates proves the bridging of the pore spaces by the solid substances. As can be seen, only a few of these particles penetrate further into the formation. If the internal filter cakes are fully developed, there will be hardly any change in the structure, even in the event of a change in the compound of the solid substance of the mud.

In due course, the solid substance particles of the drill mud get deposited on the walls of the well and build an "external filter-cake" Figure 5.

During roundtrips or the interrupt, the drilling for any other case, the mud circulation gets interrupted (static filtration).

Due to the differential pressure Δp, the liquid phase of the mud gets pressed into the formation as filtrate through the filter cakes. Apart from this, the solid substances in the cake formation get exposed to the effect of this differential pressure. With increasing thickness of the filter cake, the flow resistance of the filtrate through the cakes, acting against the differential pressure also increases.

In static filtration, the filter cake can develop further undisturbed. If the filter cake develops to the extent of equilibrium between the Δp and the flow resistance, a constant filtrate emerges.

In dynamic filtration (circulating mud), the card house structure of the clay mineral of the mud gets partly destroyed, such that a more or less oriented deposition of solid substances on the walls of the well becomes possible /103/. Due to the sheer force exerted by the by flowing mud, the development of the external filter cakes ends as soon as the friction between its outermost layers could be surmounted.

Then equations /6, 7/ apply:

$$R \leq \mu * A \left[\Delta P - \left(W \pm 2\,\sigma \frac{\cos\Theta}{r}\right)\right. \qquad \text{Eq.6}$$

$$Ks \geq R \qquad \text{Eq. 3}$$

R = the frictional force between the particles of the solid

substances in the cake formation

μ = coefficient of friction between the components of the solid substances

A = surface area of filtration

Δp = differential pressure

W = pressure loss in the pore channels of the cakes and / or the infiltration zone

σ = surface border tension between the filtrate and walls of the pore

r = radius of capillarity

Θ = angle of moisturisation

Ks = the sheer force exerted by the circulating mud of the filter cake

Through the progressive development of the filter cake, the value of W increases and this leads to a reducing frictional force between the external layers of the filter cake (compare equation 1). An equilibrium eventually emerges (see equation 7), such that the development of the filter-cake gets interrupted and a constant filtration rate emerges afterwards.

A temperature and/or an electrolyte strain of the applied clay mud affects the degree of hydration of the clay and polymer negatively. Dehydration leads to a change in the filtration characteristics of the drill mud through an excessive supply of free water as well as a change in the coefficient of friction and the nature of deposition of the clay particles in the cake formations. These changes result in intensified formation damage.

Formation damage can be through many reasons, as illustrated below:

Drilling-In

Completion

Injection

Cementing

Perforating

Hung, uncemented Slotted-Liner completion

Sand control

Production

Well stimulation

During acidizing

Kaolinit and quartz dissolution

Foam silica precipitation

Through changes in pH and salinity

Influence of chemical composition of water on clay blocking of permeability

From chemical incompatibility

Fracturing

6. DAMAGE MECHANISMS

Formation damage has been an area of concern since the early days of drilling oil and gas wells. Consequently, there has been a lot of work in the laboratory and field to learn how to reduce or eliminate damage during drilling, completion and workovers as well as damage occurring during production of the well. Similar concerns relate to injection wells. Damage during production can occur from fines (silts and clays) migrating towards the wellbore, from scale, paraffin and asphaltenes and from workover fluids. Damage during water injection can occur from formation fines being disturbed and caused to migrate, from scale (normally caused by mixing of waters, either in make-up water or commingling of waters in the formation), iron deposits (a form of scale), debris in the injected water, etc. In most vertical wells it is usually easier to find a balance between the best mud to drill through the pay zone and the amount of formation damage acceptable than in highly deviated or horizontal wells. Formation damage is usually of less concern in vertical well than a horizontal well due to a lower drilling mud-formation contact time, mud weight requirements and the ability to perforate through the damage obtained. This isn´t normally the case with horizontal wells where the majority are completed open hole or with slotted liners. Therefore, much more detail is usually paid to the mud and completion fluids used in horizontal wells from a damage standpoint than in vertical wells. In vertical wells it may be necessary to balance the mud chemistry to the most troublesome zone and thereby compromise the producing zones, whereas in horizontal wells it is more likely that the mud may be chemically balanced with the producing zone. In fact, the casing program for the horizontal well in many cases is weighted in favour of doing this. However, formation damage considerations are important in both vertical and horizontal wells. Well stimulation is about removing damage as well as increasing the ability of the reservoir to produce. To be able to remove and to stimulate reservoirs it is important to understand damage mechanisms as well as stimulation methods and materials.

Formation damage can occur in both sandstone and carbonates. In most cases damage removal in carbonates is easier than in sand stones due to their solubility in acid, i.e., we can dissolve around and remove the damage more easily with carbonates. With sandstones, the damage can be serving due to the higher probability that clay minerals will be disturbed and the difficulty in removing the resulting damage. Where clay minerals are present they are susceptible to swelling and/or migration because of the potential damage to formations containing clays considerable discussion related to clay and damage from clays is presented here. Clays are most often found associated with sandstones and shale's. Carbonate formations may also contain clays, however, these usually are encapsulated in the matrix or associated with laminate throughout the carbonate zone. With the silicates (sandstone and shale) the clays are more frequently associated with the pore space. The common clays found in sedimentary rock are Illite, Chlorite, Kaolinite, Semectite and Mixed Layer clays. Of these, the Illite, Chlorite and Kaolinite are considered non-swelling, migrating clays while Smectite and Mixed Layer clays are considered swelling clays. Analcite is a mineral of the zeolite group which is occasionally found in producing formations and is of concern since it will swell in the presence of hydrochloric acid. All the clays are made up of various combinations of two building units or sheets, the silica tetrahedral sheet Figure 6 and the aluminium octahedral sheet: By combination of these sheets, the various clay minerals are formed.

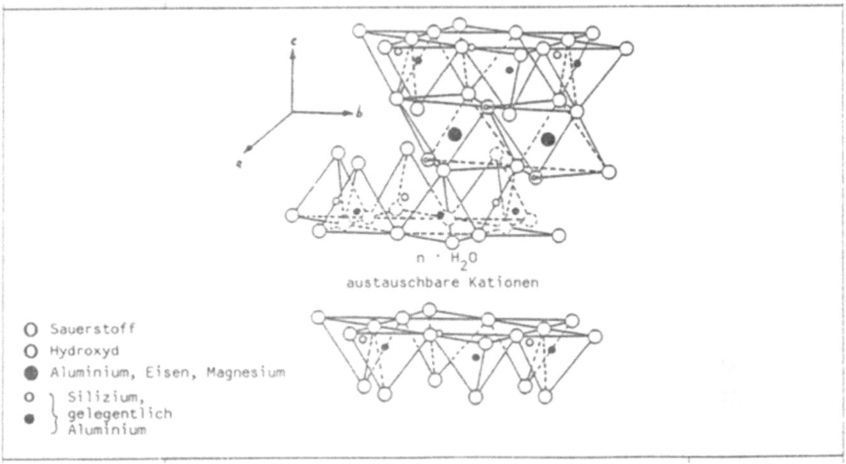

Figure 6: Aluminium Octahedral Sheet

Another example explains the Semecite of three-layer structure composed of the central alumina octahedral sheet with the silica tetrahedral sheet on each side. Substitution of Fe+2 or Mg+2 for Al+3 in the octahedral sheet and Al+3 for Si+4 in the tetrahedral sheet results in a charge unbalance of minus 1 per 1.5-unit crystals. This unbalance is neutralized by adsorption of Ca+2, Mg+2, H+, K+ and Na+ on the surface of the clay minerals. The structure of Smecite allows for cation exchange to take place within the structure. It is this cation exchange within these repeating units that allows for expansion (swelling) of Smectite clay. The mixed layer clays are composed of more than one clay mineral. Irregular mixed layer clays usually contain smectite and Illite and show similar swelling characteristics as smectite. The sheets of non-swelling clays are held more tightly together and do not swell or hydrate appreciably under normal reservoir conditions or treatment. They are, however, negatively charged on their surfaces and because of this have a cation exchange capacity. This CEC can cause these clays to be disturbed and migrate when exposed to some foreign fluids. Table 2 gives the cation exchange capacity for several clays and sand (also negatively charged) at atmospheric conditions.

Table 2: Cation Exchange Capacity for Several Clays and Sand

Clays	Range of Cation Exchange Capacity
Smectite	80-150 Meq/100g
Illite	10-40
Kaolinite	3-15
Chlorite	10-40
Sand (2-62 microns)	0.6

Where clays are associated with sandstone pores, they can be pore lining, the cementing material and/or in nodular form or any combination of these. It is not unusual to find the clays and other fine silicates held in place on the sand grains with water, the normal wetting fluid. These clays and fines can be disturbed in several ways any of which may cause migration and damage. The first of these is through cation exchange where an aqueous foreign fluid, having different ions of different concentrations, causes a change in the state of hydration of the swelling clay or a change in the association of the non-swelling clay with the sand grains and subsequent migration. Since the clays and sands are negatively charged, the clay can be held to the sand surface with cations (positive charged) in the connate water and with a cation exchange be released from this loosely held bond. The factors affecting this disturbance are:

The size of cations, the ionic diameter of cation in relation to size of the opening in the face of tetrahedral layer. This opening has been measured as being between two and three angstroms in width and depth. In addition to the diameter of the cation, the water of hydration is important in the ability of the cation to stabilize clays. As shown in the following table 3 the potassium and ammonium ion are close in size to the opening in the tetrahedral sheet (smecite and the other clays) and has a low water of hydration. Therefore, these two cations are the ones normally chosen for clay control when the control is being attempted with common salts, such as potassium chloride (KCl) or ammonium chloride (NH4Cl) /104/.

CATION IONIC DIAMETER AND WATER OF HYDRATION

Table 3: Potassium and Ammonium Ion Sizes

Cation (Angstroms)		IONIC DIAMETER HYDRATION NUMBER
Lithium	7-15	1.2
Sodium	5-11	1.90
Potassium	2.66	2.66
Ammonium	2-4	2.42
Calcium	10-22	1.98
Magnesium	12-33	1.30

The charge on the cation. Even though the multivalent cations such as calcium (Ca+2) have a stronger attraction to the clay and silicate negative charges than the monovalent cations (K+, NH+4, Na+) they have a larger water of hydration and are not as effective as the appropriate size and low water of hydration cations.

Concentration of salt (cation) shows the relative degree of smecite swelling in various salt solutions. The columns over the distilled water caption show how much swelling could be expected in distilled water if the smectite clay has sodium, potassium or calcium associated with its structure. The next other three columns for 0.4% sodium, potassium and calcium chloride show that at this concentration the potassium chloride allows less water to be absorbed on the smectite clay implying it is more effective in controlling clays. At 2% chloride concentration the potassium chloride is still more effective than sodium or calcium chloride. At 8.5% concentration the sodium and potassium chloride are about equal. This data implies that approximately 0,4% potassium chloride is equivalent to 8,5% sodium chloride and that 0,4% potassium chloride may be all is necessary for clay control. Two points are important here, one is that as fluid leaks off into the formation the potassium ion is pulled from the solution and it becomes less potassium concentrated with time for this reason 2% KCl is the normally accepted minimum concentration to provide clay control using salt. The second point is that if at least 8.5% sodium chloride is used it would probably not be advantageous to add potassium chloride for additional clay control, based upon this information for smecitite. The type and concentration of clays may dictate different amounts of salt in solution for best clay control using this method. Laboratory testing is recommended to evaluate the best materials and concentrations for a given formation /20/.

pH of water contacting clays is important. pH is a measure of the hydrogen ion concentration, pH = - log (H+). A pH of less than 7.0 indicates an excess of H+, and an acidic solution. This hydrogen ion behaves as other cations in assisting clay control. Conversely, if the pH is greater than 7.0 there is an excess of hydroxyl ions (OH-), and a basic solution. The hydroxyl ion aggravates the clay hydration and migration problem. Therefore, more clay disruption is anticipated in a high pH fluid as compared to low pH, <7.0, fluids. When bentonitc (smectitic) water base drilling muds are used it is common to maintain the pH at > 8.0 to assist in viscosity yield (bentonite swelling), for corrosion and bacteria control. This increased hydration (yield) of the bentonite from high is the same increase in hydration we would expect from swelling clays in the formation. Additionally, we would expect more disturbances of the non-swelling clays in high pH fluids due to the repulsive forces between the clay surface and the hydroxyl ion (negative charges). Usual potassium concentrations do not appear to override this disruptive effect of high pH.

Type, location and amount of clays, obviously all of the above 5 factors are affected by the type and amount of clay present in the formation. Additionally, the distribution of these clays, pore lining, nodular, etc. will influence the damage experienced.

Cationic organic polymers (COP) are used for clay control and can be more effective than the salts in controlling clay movement in the formation. These polymers are highly charged (+) and tend to blanket the cation exchange sites of the clays and offer long term protection. They are used in many completion fluids but are most popular in fracturing and acidizing treatments.

The three common laboratory methods to analyze formation samples for potential damage with various fluids are X-ray diffraction, thin section – petrographic analysis and scanning electron microscope, (SEM) which I used in my experiments: In additional to these cores, flow through tests is often conducted. X-ray diffraction can show what crystalline materials, minerals, are present. This analysis is usually a twostep process where a sample of the formation is ground to a powder consistency (bulk sample) and a clay fraction is separated from the bulk sample by water suspension and allowed to dry on a glass slide. This concentration of the small particles and clays on the glass slide allows a good X-ray diffraction analysis of the clays in the sample. The bulk sample is analysed by X-ray diffraction for the non-clay minerals. Thin section analysis utilizing a petrographic microscope allows observation of the various minerals in relation to the pore space. This analysis is an excellent way to see what minerals invading fluids will contact. Scanning electron microscope (SEM) analysis allows observation inside the pore space and is best utilized on a before and after fluid contact test. This type of analysis shows what detrimental effect certain fluids may have had after contact with the pores and formation minerals.

Additional analyses can be done on whole core sample where flow test are conducted to determine sensitivity of the formation to various fluids as they are flowed through the pore spaces, which I examined in the research programs. These tests usually include regained permeability to a produced fluid after contact with test fluids. Comparative analyses help to further determine fluid – rock compatibility. These tests should always consider the fluids in the formation, including mud filtrate, etc. These analyse and concepts help us understand if significant damage may occur with the various fluids which may be put on the formation, when considering the drilling mud for the horizontal well, the completion fluids and workover fluids.

Another damage mechanism which should be considered is change in wettability. As was seen in Figure 1, a change from a water wet to an oil wet rock can drastically reduce the relative permeability to oil. This change from water wet to oil wet can occur with some drilling muds, naturally occurring surfactants in some oils, surface treated fluids, etc. To maintain water wet conditions in sandstones or shale's anionic or non-ionic surfactants should be used. With carbonates, cationic or non-ionic surfactants should be used to maintain water wet conditions. Non-ionic surfactants can be effective in silicates and carbonates to maintain water wetting. However, they should be evaluated in the laboratory to determine their applicability. Asphaltenes in the formation oil can cause formation plugging as well as oil wetting of sandstones and shale's if they come out of solution. They can be precipitated from oil by contact with straight chain hydrocarbons, such as diesel or kerosene, and with acid and possibly other mechanisms. Surfactants are important additives used in various fluids in drilling and well completions. An understanding of their basic properties can help in prevent or overcoming formation damage. Surfactants can be characterized as having a water-soluble head (hydrophilic) and an oil soluble tail (lipophilic). They are further defined as water soluble or oil soluble based upon the relative length of the oil soluble tail. An example of water and oil soluble surfactants is shown below.

Surfactants are further defined by the charge at the water-soluble end. This charge relates to the wetting characteristics on various types of formations (A, B, C, D).

A

+ 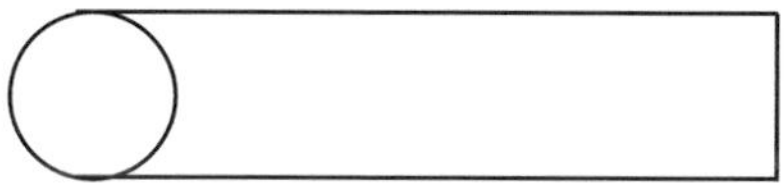

a) Cations – Oil wet sandstone, shale's clays and other silicates. Water wet carbonates at pH less than 8.5

B

- 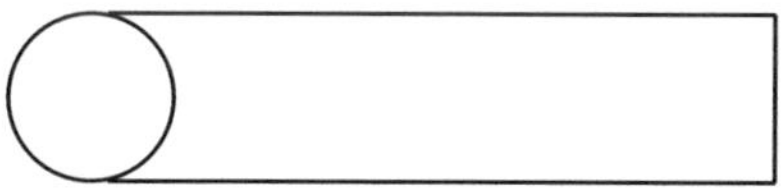

b) Anionics–Water wet sandstones, shale clays and other silicates. Oil wet carbonates at pH less than 8.5

C

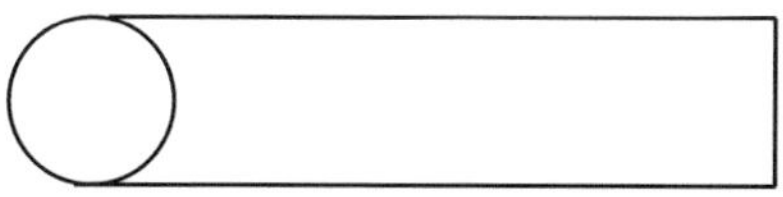

c) Non-Ionics – Wetting characteristics depend upon the relative length of the oil soluble tail. Can water wet or oil wet silicates and carbonates depending upon the surfactant chemistry.

D

+ / - 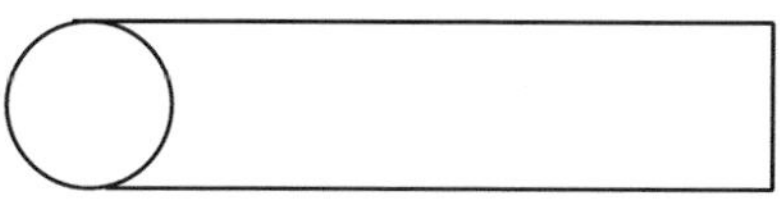

Amphoteric – This surfactant can be either or anionic depending upon the pH of the fluid it is in. Therefore, wetting characteristics will vary.

The wetting characteristic for the cationic and anionic surfactants relate to the charge on the silicates and carbonates. The silicates have a negative charge on their surface and carbonate a positive charge at pH less than approximately 8.5.

When a surfactant is attracted to pore space surfaces as with cationic on sandstone and anionics on carbonates, oil wetting can occur. Several negative effects result from this.

a) The formation is oil wet and relative permeability to oil greatly reduced

b) Surfactant is lost from solution and no longer available to assist in fluid recovery via reduced surface/interfacial tension

c) Surfactant is lost from solution and no longer available to prevent emulsions from occurring

Other sources of damage while drilling, completing, doing workovers and later during the production of the well, are migration of fine particles in the reservoir reducing near wellbore permeability, the formation of emulsions, precipitation of paraffin and asphaltenes and scale deposition. The fine particles which may migrate can be clay, silt and other small particles which are normally either in the pore space or adhering to sand grains. The small particles adhering to sand grains are normally associated with the connate water coating each grain. As long as the wetting phase (water) is not flowing or the water layer on the sand grains encapsulating the fines is not disturbed these particles can be relatively stable. However, when the wetting phase begins to flow or production fluid velocity is sufficiently high these particles can begin to move and cause plugging. Any of the factors discussed earlier which disturb the clays can aggravate the migration of clays and other fines.

Emulsions are caused when two immiscible fluids come in contact with each other in the presence of an energy source and when stabilizers are present. The most significant emulsion stabilizers are fine particles of clay or other materials when oil wet, asphaltenes and surfactants. The surfactants causing emulsions can occur naturally in crude oil or may be present in treating fluids. Emulsions are defined as droplets of one fluid suspended in another. They are either oil or water external. Oil external emulsions are those with oil as the continuous phase and water, acid, etc as individual droplets suspended in the oil. Water external emulsions are those with water, acid, etc as the continuous phase and oil droplets suspended in the water. The viscosity of emulsions is controlled by the ratio of oil and water making up the emulsion, the viscosity of external phase and the amount of energy put into the emulsion at the time it is formed. Oil external emulsions are normally thicker and more viscous than water external emulsions, primarily because the oil usually has a higher viscosity than water. Obviously, this is not always the case and the other two factors mentioned above influence the viscosity of the emulsion. Regardless of which type emulsion exists, they can have a very damaging effect on production. The type of emulsion, water or oil external, can determined easily by dropping some of the emulsion in water and watching for dispersion. Water external emulsions, when placed in water allow the oil droplets to disperse while oil external emulsions dropped in water allow no dispersion. Emulsion prevention and removal is normally handled with surfactants that counter the effect of naturally occurring surfactants, asphaltenes, and fines that have caused or may cause emulsions. Surfactants are added to stimulation fluids, workover fluids, etc to prevent emulsions from occurring during and after the treatments, for fluid recovery and to maintain wettability and should be chosen based upon laboratory testing. Not only is the selection of the proper surfactant important but the concentration of the surfactant is important. Over treating can actually cause an emulsion when the objective was to prevent emulsions. When treating for emulsions it is best to use the same fluid type as the external phase of the emulsion, i.e., treat water external emulsions with water base fluids containing the appropriate surfactant and oil external surfactants with oil base fluids containing the appropriate surfactant. Figure 4-5 shows how surfactants, naturally occurring or added to treating fluids, orient themselves at interfaces and can stabilize emulsions. Emulsions can be difficult to treat due to our inability to contact all of the emulsion. Additionally, the completion may not allow adequate placement of the treating fluid. This is especially true in many horizontal wells.

Paraffin and asphaltenes deposited in wells can be equally difficult to treat. A good understanding of the crude oil characteristics can help in controlling the deposition of either these materials during drilling, completing, workovers or production. Where the problem is server the completion program should be designed so the wellbore can be treated as necessary to prevent production problems. Paraffin's and asphaltenes differ significantly in chemical structure and should be considered as two separate organic materials, even though they may both be found in an oil and deposited together at times. Paraffin's are normal (straight chain) or branched alkenes of relative high molecular weight (C18 to C70) and are represented by the general formula Cn H2n+2. Each specific paraffin has a specific melting point, e.g., melts at 82 °F (27.8°C) and C60 melts at 215 °F (101.7 °C), and specific solubilities in various solvents. Most paraffin deposits are mixtures of these alkenes and consist of very small crystals that usually agglomerate to from granular particles with varying melting temperatures. The term wax usually refers to solid paraffin and is often used interchangeably with the term paraffin. Paraffin's are soluble in most petroleum solvents, both straight chain and aromatics. Paraffin loses its solubility in crude oils primarily due to cooling. The temperature at which paraffin begins to come out of solution can be determined by an ASTM D 2500-66 testing procedure for cloud point, the point at which the paraffin crystals begin to come out of solution. In light-coloured crudes, as temperature is reduced and paraffin begins to come out of solution, the crude takes a cloudy appearance. The temperature at which this happens is known as **the could** point. Further cooling of the crude results in a pour point, the temperature at which the crude can no longer be poured from the test container. The cloud point temperature is the temperature at which paraffin deposition can begin in a production system, from the reservoir to surface production facilities. When crude oils are dark and it is difficult to see the cloud point in a test cell, a measure of viscosity versus temperature can show cloud point by a change in the viscosity at the time paraffin crystals begin to come out of solution. In light-coloured crudes, as temperature is reduced and paraffin begins to come out of solution, the crude takes on a cloudy appearance. The temperature at which this happens is known as the cloud point. Sources of cooling are drilling, completion and workover fluids, gas coming out of solution and reduction in production rates (allowing cooling of the production string) When paraffin crystals begin coming out of solution they have a needle-like shape and are not necessarily a problem until they begin to agglomerate This agglomeration is accelerated by nucleating agents such as fines, asphaltenes, scales, etc.

Asphaltenes are black, polycyclic aromatic, complex compounds and have a much higher molecular weight than paraffin. They are surrounded by lower molecular weight resins and aromatic hydrocarbons asphaltenes occur as dispersed solid particles in crude oil and sometimes in condensate associated with gas production Their polar, electrically charged, properties result from the presence of oxygen, sulphur, nitrogen and various metals in their structure Due to this polar characteristic, when asphaltenes come out of solution they are cationic in nature and will tend to oil wet sandstones as well as being a solid material which can be pore blocking /105/. Thus, asphaltene deposition in the near wellbore area can have a dramatic effect on production asphaltene deposition usually results from a drop in pressure or destabilizing factors such as contact with some acids, Emulsions are caused when two immiscible fluids come in contact with each other in the presence of an energy source and when stabilizers are present. The most significant emulsion stabilizers are fine particles of clay or other materials when oil wet, asphaltenes and surfactants. The surfactants causing emulsions can occur naturally in crude oil or may be present in treating fluids. Emulsions are defined as droplets of one fluid suspended in another. They are either oil or water external. Oil external emulsions are those with oil as the continuous phase and water, acid, etc as individual droplets suspended in the oil. Water external emulsions are those with water, acid, etc as the continuous phase and oil droplets suspended in the water. The viscosity of emulsions is controlled by the ratio of oil and water making up the emulsion, the viscosity of external phase and the amount of energy put into the emulsion at the time it is formed. Oil external emulsions are normally thicker and more viscous than water external emulsions, primarily because the oil usually has a higher viscosity than water. Obviously, this is not always the case and the other two factors mentioned above influence the viscosity of emulsion. Regardless of which type emulsion exists, they can have a very damaging effect on production.

7. CAUSES OF FORMATION DAMAGE

Formation damage is also caused through an invasion of substances (liquids, solid substances) that are foreign to the reservoirs and/or a further migration of the contents of the reservoirs within the reservoir.

The infiltration in the formation, that can partly contribute decisively to formation damage, depending on the depth of penetration, takes place under a total pressure Pg:

$$Pg = \Delta P \pm Po + Pc \qquad \text{Eq. 4}$$

$$Po = c \times R \times T \qquad \text{Eq. 5}$$

$$\mathbf{PC} = \frac{\mathbf{2.\ \alpha.\ \cos\theta}}{\mathbf{r}} \qquad \text{Eq. 10}$$

ΔP = Difference between the pressure of the mud-pile and the pressure of the pore-content´s.

Po = Osmosis pressure

c = Difference in ion concentration between the mud and the pore content

R = Universal gas constant

T = Temperature

Pc = Capillarity pressure

σ = Boarder surface tension

r = Radius of capillarity

Equation 8 makes it clear that even in "balance drilling" itself, (ΔP= 0) loss of filtrates are possible in the formation.

In view of the causes one differentiates between:

- Physically,
- Chemically and
- Bacterially

Physically Caused Formation Damage

During the Mud Spurt-Phase, solid substances penetrate undisturbed into the pores of the formation rocks and become entangled in the necks of the pores. This leads to a reduction of the cross-sectional surface of the capillaries, which results in a reduction of permeability.

After the development of an internal and/or external filter cakes, the filtrate then penetrates only into the formation. The change of the original saturation caused through this can reduce the relative permeability for the phases to be produced. Both phenomena are combined together under "physically caused formation damage"

Chemically Caused Formation Damage

The chemically caused formation damage also takes place through the mud filtrate that gets penetrated into the formation. Through swelling (expansion) and the dispersing of the reservoir clay respectively, break-offs of particles from the matrix as well as fall-outs of insoluble salts from the reaction with the substances of the pore contents, the filtrate can cause a high reduction in the permeability of the formation.

CIVAN, KNAPP and OHEN /108-110/ proved that the main cause of formation damage is the increase in volume of the clays of the reservoir. Depending on its convalescence, the swelling (expansion) of the clay takes place in the pore spaces and the on the pore walling respectively.

According to VADYA and FOLGER /116/ the "in-situ"-loosening of clays and other components of the matrix is conditioned above all through exchange of ions between the filtrate and the rock.

As a rule, sandstone contains clays that would either primarily be deposited during the sedimentation (detritic clays) or secondarily be stored through the transportation in the pore spaces by the diagens (digenetic clay).

Through the increment of the degree of the hydration of this clay, the filtrate that penetrates into the pore spaces can cause formation damage; in this case, the level of the hydrations grade depends on the difference in ion-concentration between the filtrate and the pore contents as well as the portion of free water in the mud.

Through the taking up of filtrate, the clay can swell (expand) and cause the contraction of the capillaries; it is also possible that the deposited clay gets dissolved and transported further by the filtrate, such that these get caught in the narrow parts of the capillaries and cause obstructions.

To rule out a formation damage through swelling (expansion) and dispersing of its clays respectively, provision must be made for the maintenance of the electro-chemical equilibrium in the pore spaces /4, 5, 18, 24, 50, 121-126, 136/. The realisation is possible through the appropriate salting of the liquid phase of the mud and the binding of the free water through polymers.

If the rock matrix contains soluble components, the filtrate can cause a weakness of the matrix through the dissolution of such components.

Alongside, matrix components can be freely laid, and in the course of further transportation and settlement in the narrow parts of pore channels, this could cause formation damage.

Damage through Bacteria's

A multitude of micro-organisms are often found in water-based muds /137-146/. Their growth in the pore spaces, their remnant's and wastes can immensely lower the permeability in formations with low permeability /117, 138, 139/.

By the supply of nutrients in the reservoir through the filtrate, these bacteria's grow and surround themselves with self-produced polysaccharides, which considerably reduces the permeability heavily, as would be the case in a pure cell growth /140-143/.

Damage due to Drilling

Some damage during drilling is to be expected. This damage will be from the mud filtrate and from particulate material in the mud, drill solids as well as weighting materials. The filtrate from the mud can be the most damaging when formation clays are present and the mud has not been designed to protect from this damage. This filtrate can contain oil wetting materials used in the mud. However, oil wetting producing formations should be avoided if practical it is possible that an oil base mud is necessary to be able to drill through some formations and there really is no other option but to accept the damage from oil wetting. The oil wetting can be corrected during the completion of the well if proper precautions are taken. Emulsions can form when mud filtrate mixes with formation oil and when certain surfactants used in some muds interact with connate water and formation crude precipitation of scales, paraffin or asphaltenes can also result for contact of incompatible mud filtrate and formation fluids. Invasion of mud solids into the formation can occur prior to or during the build-up of a filter cake. Factors influencing the invasion of mud solids into the formation include size of the formation pores, size and concentration of the solids in the mud, filter cake build-up rate, mud weight in relation to the formation pore pressure, drilling penetration rate and the circulation rate of the mud. Works presented in the past has indicated that if the solids are less than the size of the pores then the solids can invade the pores. An estimation of pore size can be made if the permeability is known, as shown below:

Where:

dp = pore throat diameter, microns

k = formation permeability, millidarcy

With this very general rule of thumb, we can estimate the particle size that could enter the formation. It can also be used to determine fluid loss particle size to assist in controlling fluid loss. In many cases fluid loss additive sizes are referred to sieve sizes. These figures can help in converting these to the size units used for the pore's particle size are critical to the formation of a good filter cake. A mud containing all solids smaller than the pore size will not readily form a filter cake. A mud containing all solids greater than pore size will quickly form a filter cake.

Damage due to Cementing

Damage to productive intervals can occur during the cementing operation. Prior to running casing the hole is normally conditioned with the mud to clean the hole of drill solids, etc. The casing is then run and cemented.

If all the filter cake and mud solids are not removed, fluid loss from the spacers may be minimal. More fluid loss will occur when a spacer does a good job of cleaning the hole. Additional fluid loss will occur if the pipe is reciprocated and or rotated. This will accelerate the mud cleanout and displacement as well as increasing the fluid leak off. Moving the pipe is usually recommended for good cementing results. Whole cement particles do not usually invade the formation. However, recently developed microfine cement is designed to enter formation pores but is usually only used for remedial cementing work and not primary cementing. Normal cement particles range in size from 20 to 100 microns, which are too large to fit in most formation pores. The primary potential formation damage from cementing is from the water portion of the cement. The pH of the water after having been used to prepare the cement is very high, >10, and can cause problems in those formations containing clays. Properly designed and mixed cements should lose a very small amount of the mix water. If there is enough water lost from the cement to cause formation damage then the cement was either mixed with excess water and or the fluid loss was excessive. With excess fluid loss a very poor cement job would result with probable dehydration and difficulty in circulating the slurry. For these reasons, formation damage from primary cementing should not be a problem if proper procedures are followed. In those cases where lost circulation has occurred in the target zone, whole mud and cement can be lost and these create formation damage problems which are difficult to overcome. Because of the volumes lost, much of the near wellbore area is affected and significant damage can result. When a poor primary cement job is carried out communication may exist between the pay zone and adjacent zones. This communication can cause damage in several ways. If there is communication with water and it is incompatible with the pay zone, either due to clays, precipitates, emulsions, etc, a physical damage would result. Due to the water invasion into the pay zone, the relative permeability will be changed and perhaps only water will be produced. Where communication is from a gas to an oil producing zone a high GOR may result. So, good cementing practices can go a long way in keeping formation damage from occurring during this phase of work. Where remedial squeeze work is necessary, obviously (extreme over balance) of perforating. These treatments include a fluid covering the gun and, in most cases, additional fluid is pumped immediately after the guns are fired. The intent is for the high pressure at the formation to fracture the formation around each perforation at the time the guns are fired. The additional fluid pumped would normally contain a propping agent if sandstones are being perforated and possibly an acid if carbonates are being perforated. This concept can result in overcoming damage and possibly stimulation following the perforating operation. Again, careful selection of the fluids used would be important to assure the planned outcome. Most perforating is done with jet charges which push pipe, cement and formation material out in front of the jet penetration. This type of perforating results in a compacted zone around the perforation tunnel and a reduction in permeability. The degree of this permeability reduction is related to the perforating charge, rock properties and the wellbore pressure at the time the guns were fired.

The main causes of formation damage are:

Formation Damage through Solid Substances

The pressing of formations through solid substances can occur on the walling of the borehole, along the filtration routes and in the formation. In this connection, burdensome materials clays, drill particles, polymers, bridging substances, cement particles, paraffin's asphalts and insoluble salts are solid substances that come into question.

Bigger substances can get settled on the surface of the rock and lead to the building of an external filter cake. Cracks and bigger pore channels can also be bridged through these solid substances. Smaller particles can penetrate relatively deep into the formation; the development of flow barriers within the formation can occur.

By the take-off of exploitation, a dragging along of fine particles can occur through increased flow velocity, which gets caught in the narrow parts of the capillaries Figure 7.

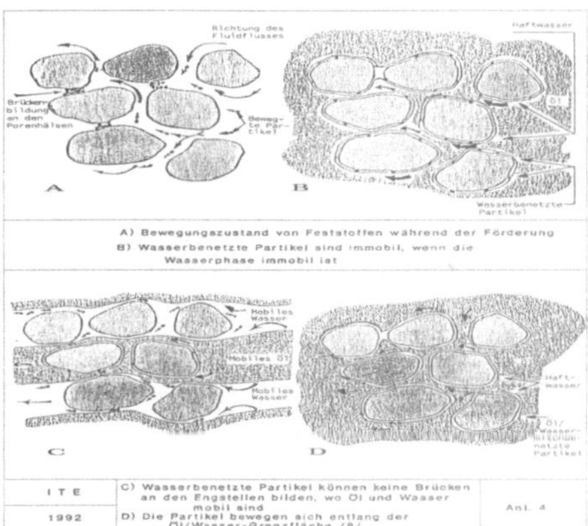

Figure 7: Fine Particles Dragging along the Capillaries

Clay particles can be reduced through the treatment of the formation with hydrochloric acid, in which the H -ions of the acid replace the exchange able cations of the clay; the hydrate coverings of the clay particles will be partly dismantled through this substitution. The permeability of the formation can be increased through this treatment /101, 102/.

Due to its deformability, hydrated polymers can be washed extensively into the formation. The hydrated polymers and the pore walling of the rock normally show different electrical charges, which lead to the absorption of the polymer; the result is the contracting of the capillaries.

Summarily, the following mechanism can be held responsible for the blockage of the flow routes through an invasion /8, 93-98,100-118/:

a) sticking of the solid substances on the walls of the pores through electrostatic, hydrodynamic or gravitational forces

b) settlement of the washed in particles in the narrow capillaries

c) building of bridges through particles that are smaller than the openings of the pores of the rocks

Building of bridges can occur when the size of the particles of the bridge building materials amounts to ca. 1/3 of the openings of the pores of the reservoir rock /4-6, 8, 13, 15, 16, 18, 22, 55, 62, 64, 71/.

Formation Damage through Fluids

In deep drillings, different sorts of fluids with different concentration of ions penetrate as filtrate into the pores of the rock; they can displace the original phases in the pore spaces (changing of the ratio of saturation) and/or get mixed with the contents of the pores (among others building of emulsions). Besides, a reduction of the permeability of the formation can occur for the phases to be produced. Factors that decrease the productivity of a reservoir through the invasion of filtrate are /1, 3, 46, 13, 15, 16, 18-21, 32-35, 43, 45-49, 51, 59, 62, 64, 71,72, 78, 113, 115 /:

a) Reduction of the relative permeability for the oil and/or gas phases, caused through increased water saturation (water block)

b) Reduction of the absolute permeability of the formation rock through narrowing of the pore channels and

c) Increase of the viscosity of the oil-phase through the building of emulsion

If the rock matrix in the filtration zone is to be oil-moistured through active surface boarder substances, bactericide, corrosion protective medium, jel-breaker or oil-based muds, the produced oil phase can cover the walls of the pores with an oil film, which lead to a narrowing of the flow channels of the formation rock /8,76/.

The filtrate of water-based muds displaces the oil in the reservoir and consequently changes the saturation ratio in the pore space.

At the beginning of production, the conveying flow of the filtrate must then be repressed, for which a critical pressure, the so called "threshold pressure", must be subdued. The phenomenon described is referred to as "water block" /4, 5, 16, 18, 71, 76, 119/.

A mixture of oil and filtrate can lead to the building of a stable emulsion in the pore space (emulsion block). Generally, the crude oil contains resins and asphalts. The latter are alkaline salts of organic acids. Their molecules are wrapped up in hydrocarbon molecules. The hydrocarbon covers irreversibly burst open during the transportation through the pore channels. The resins and asphalts get concentrated on the surface boarders of the oil filtrate and consequently cater for the stability of the emulsion /4, 5, 43, 51, 71, 72, 76, 89-91, 120/.

Depending on the characteristics of the emulsion and the moisturisation of the formation, oil/water membranes from adsorbed, diaphragmental materials get build up on the boarders of the layer. A reinforcement of this film of the surface boarder happens through fine portions, asphalts and increased salt content.

Chemical reactions between the filtrate and the contents of the pores and the rock matrix respectively can lead to a gas development.

This additional phase reduces the relative permeability's for the remaining phases in the pores ("gas block").

Formation Damage through Treatment Liquids

Frac Fluids: Figure 8 shows parts that may likely be damaged in a Frac treatment /29,30/ Breakings and the embedding of the supporting material could occur within the cracks respectively, if this is selected without considering conditions of application. Further causes of damages are the dissolution of the supporting material through hot formation waters and their transportation back into the borehole respectively during the free exploration phase. Eventually, the dissolution of the matrix material through the further transportation of the released solid substances can lead to the blocking of the flow route /106/.

The amount of damage in the case of a Frac treatment depends on the type and concentration of the polymers, effect mechanism of the gel breaker, the temperature and the pH value.

The natural products Guar-gum, Hydroxypropylguar (HPG) and Xanthan contains certain quantities of insoluble portions, while the cellulose derivatives are practically fully dissociable /25, 73-85/.

Corrosion's residues from treatment tanks and pipes can make up to considerable volumes and contribute to formation damage /76, 86/.

Depending on the quality, the supporting material can also contain up to 1% weight of clay or finer stone particles.

In Frac- or Gravel-treatment respectively, the penetration of fluids take place, this can lead to precipitations in the depth of the formation through the physical-chemical equilibrium between the pore contents the treatment liquid. Then in the case of water-based liquids, this has to do with precipitated carbonates, sulphates and oxides.

In the application of oil-based fluids, asphalts and paraffin's can precipitate (fall-out) and get deposited in the pore spaces. Therefore, the compatibility of the treatment medium with the contents of the reservoir must be investigated before the stimulation /20, 51, 75-78, 81-86/.

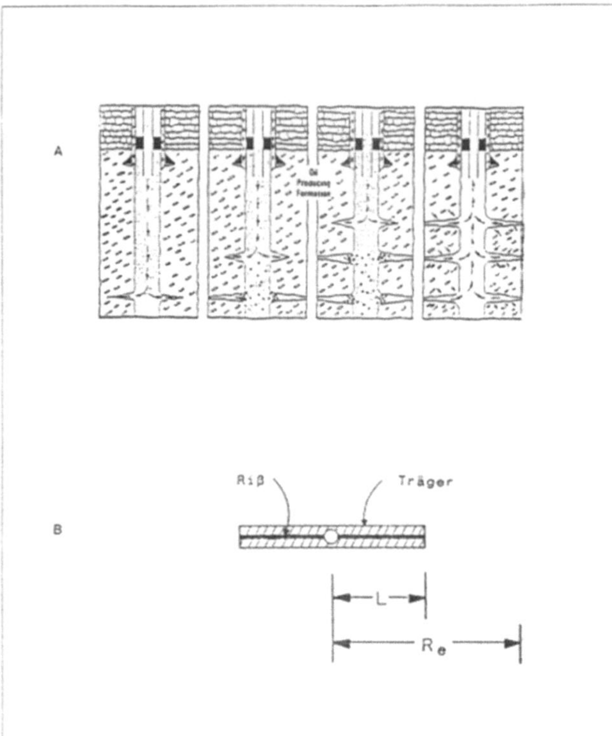

Figure 8: Range of possible damage after Frac treatment /29, 30/

Completion Fluids: Completion and workover fluids are used for their ability to not only control formation pressure, but also to reduce/eliminate certain types of formation damage. The two basic types of completion and workover systems are clear fluid systems and solids enhanced systems.

In some fields, an only clear fluid filtered system is used. When determining if a fluid will perform effectively in the planned completion or workover operation, it is necessary to consider the following factors:

Density

Crystallization point

Brine/formation water compatibility

Corrosion

Clear brines will be used in both underbalanced and overbalanced conditions. Frequently, a well is completed in an overbalanced situation and the heavy brine is replaced with a lighter packer fluid.

If a well is completed...	Then...
• In an underbalanced situation, •With an underbalanced packer fluid left in the well,	Casing design and cost are the main factors to consider when selecting a brine density and corresponding brine.
• In an overbalanced situation, • As a workover operation requiring a kill fluid,	The required density is determined by formation pressure, true vertical depth, and temperature gradient.

In the case, perforations will be done in an overbalanced situation. Brine's crystallization point is the temperature at which salt crystals will begin to fall out of solution given stime and proper nucleating conditions.

The precipitation of insoluble salts can cause a number of problems. For example, when the dissolved salt in the fluid crystallizes and settles in a tank, the fluid density usually drops. Crystallization in brines can also cause lines to plug and pumps to seize /130-135/.

To ensure crystallization does not occur in the brine:

a) Determine the required crystallization point of the fluid

b) Check the actual crystallization point of the fluid

c) Adjust the crystallization point of the fluid, as necessary

d) Use heavier components if saturation is likely to occur

To select the correct brine type, consider the potential interactions of the completion or workover fluid with formation solids, water, and gases. Most common incompatibility problems include:

a) Scale production from the reaction of a divalent brine with dissolved carbon dioxide

b) Precipitation of sodium chloride from the formation water when it is exposed to certain brines

c) Precipitation of iron compounds in the formation resulting from interaction with soluble iron in the completion fluid

d) Reaction of formation clays with the clear brine

The following laboratory tests can be used to evaluate the compatibility of a clear fluid with a formation:

a) Return permeability

b) Formation water analysis

c) Formation mineralogy

d) Brine/water compatibility

e) Rock sensitivity

All these tests will be done during the development of the project.

The corrosivity of a completion or workover fluid depends on its type. Monovalent fluids generally show low corrosivity, even at temperatures exceeding 400°F (204°C). The corrosivity of divalent fluids depends on the density and chemical composition of the fluid. Laboratory data show that, for divalent fluids not treated with corrosion inhibitors, the addition of calcium chloride gives a slower rate of corrosion compared to that of zinc bromide which gives a faster rate of corrosion.

Contaminants that can affect completion and workover fluids include:

a) Iron
b) Solids
c) Hardness
d) Oil, distillate, grease, and pipe dope
e) Polymers
f) Surfactants

Iron can be a contaminant in either soluble or insoluble form. Soluble iron is a product of corrosion and is common in zinc fluids. When exposed to certain waters, soluble iron can form a precipitate, which can cause formation damage. The soluble iron content of a fluid can be measured using the iron test kit normally in the field test available.

At the brine plant, iron should be removed from a fluid by adding hydrogen peroxide to the fluid, flocculating the fluid, and then filtering the fluid. On location, treating a fluid for iron is very difficult and is usually successful only in low-density brines such as KCl, NaCl, or CaCl2. The treatment consists of increasing pH with caustic or lime and removing the precipitated iron by filtering the brine.

Total solids can be measured at the wellsite using a turbidity meter (Field tests) or a shake-out machine. Solids that are not added to the system to enhance the performance of brine are considered contaminants. Contaminants include formation clays, precipitates, and polymer residues, among other things. These contaminants can be filtered at the wellsite using diatomaceous earth, a plate a press, and two-micron absolute cartridges.

When a monovalent brine has been selected to minimize calcium and magnesium scale formation, the total hardness content should not exceed 100 mg/L. Brines contaminated in the plant should be treated with soda ash and/or BARASCAV and filtered. To settle the precipitate prior to filtration, a flocculant might be needed.

Produced oils and other hydrocarbons affect brine density and can also blind filtration units. Hydrocarbons will form a separate layer above heavy brine and should be pumped off the surface.

Brines contaminated with polymers usually cannot be filtered without chemical and/or special mechanical treatment at the plant site where hydrogen peroxide can be used to oxidize polymers and permit filtration. At the wellsite, polymer pills used in displacement should be caught and isolated from the active brine system. Compatibility tests and formation damage tests should be run with any surfactant package required for completion.

8. EVALUATION OF DAMAGE

Prior to any corrective action, an evaluation of damage and inflow performance should be carried out Production logs can provide excellent information about the productivity of the various intervals which should be providing production. Questions which can be addressed by production logs are: a) mechanical conditions of the well, i.e., casing, tubing or packer leaks, internal or external corrosion, b) anomalous fluid movement between zones, i.e., flow behind casing due to poor primary cementing, crossflow inside casing, c) completion efficiency evaluation of producing wells, i.e., are all zones producing, are some zones producing only gas or water, are the zones producing up to their potential indicated by other data sources d) completion efficiency evaluation of injection wells, i.e., how much fluid is going into various zones, e) design and evaluation of stimulation treatments, i.e., which zones need stimulating, which zones responded to stimulation f) reservoir management, i.e., what are initial fluid saturations in each zone, what fluid saturation changes have occurred since initial completion, is the reservoir being depleted in the desired manner. To add confidence to downhole analysis usually more than one logging device or measurement is required. The types of measurements which can be made downhole are:

Temperature, which is a very basic measurement but can provide valuable information about fluid movement downhole, both inside and outside the casing. Temperature inside the tubing or casing is a result of heat flow and for fluid expansion. Heat is transferred through materials and liquids by conduction influenced by temperature difference. Heat is also transferred by convection due to fluid movement influenced by flow rate, temperature differences and fluid type. Temperature surveys following injection of fluids for water must be reasonably low. Also, fluid saturations in the rock near the wellbore may not be the same as those further away. Pulsed neutron bombardment of elements such as carbon, oxygen, chlorine, silicon, hydrogen, calcium, sulphur and iron in the rock near the wellbore permits the measurement of the relative saturation of the fluids with which they are associated. Many factors enter info calculation of fluid saturation by pulsed neutron techniques; thus, these devices are most effective when used to compare changes in fluid saturation over a relatively long period of time due to production or a short period of time due to injection. Audible noise level and frequency patterns in the wellbore caused by movement of fluids inside or outside the casing can be used to establish the presence of flow, the path of flow, what phases are involved and to a degree, the flow rate. Sound transmission characteristics from a wellbore transmitter to a nearby receiver provide information as to acoustic coupling between the cement and the casing and the cement and the formation. Under favourable conditions these characteristics may be related to the possibility of fluid movement between zones outside the casing

/23/, electrical properties (i e, conductivity or dielectric constant) usually differ significantly between hydrocarbons and water. Thus, they may be used to determine relative amounts of these fluids at a particular level in the wellbore. These measurements are more definitive at low water percentages and where the fluids are intimately mixed again, the relative amounts of fluids residing at a particular level are not necessarily the same as the relative amounts of the fluids passing that level. An excellent measure of a well performance is through well testing. Oil and gas well production tests are generally classified as a) periodic production tests, b) productivity or deliverability tests and c) transient pressure tests /9/. Periodic production tests are run to physically measure oil, gas and water produced by a particular well under normal producing conditions. Test results may be used to allocate total field or Specific PI accounts for the length of the producing section as follows:

$$\mathbf{PI} = \mathbf{J} = \frac{\mathbf{Q}}{\mathbf{(Pi - Pwf)}} \qquad \text{Eq. 6}$$

Where:

PI = Productivity Index /Length of Producing Zone

q = total liquids stb/d

pi = shut-in bottomhole pressure, psi

pwf = flowing bottomhole pressure, psi

pi - pwf = pressure drawdown, psi

When the well producing above the bubble point the PI may be constant over a wide range of pressure drawdown's However, with the flow below bubble point, and gas occupying a portion of the pore system, PI falls off with increased drawdown. Productivity index also declines during the life of a well due to many factors, among which are changes in reservoir pressure, composition and properties of produced fluids, relative permeability change and flow restrictions or formation damage near the wellbore. Productivity index does, however, give us a useful index of well and wellbore conditions. As long as we recognize the limitations of this method of well evaluation it can give us a measure for comparison between wells. Inflow performance testing is an improvement over the PI approach since it, in essence, looks at PI´s at several production rates PI attempts to represent the inflow performance relation of a well as a straight line function. The true inflow performance relation or IPR usually declines at greater drawdown's, as shown in through the surface equipment. Each of these nodes can result in a pressure drop which the reservoir pressure much work against to produce various relationships are available to determine these pressure drops for various fluids being produced. This nodal system analysis will help in determining which nodes are limiting production and help in separating the difference between true formation damage and pseudo-damage, i e., completion restrictions. Flow after flow testing has been used for many years for determination of gas well capability. Essentially a plot of flow rate versus squared drawdown pressure on log-log paper provides a straight line which can be extended to predict flow rate for any drawdown. Oil well deliverability can be represented in the same manner. The method is particularly used for reservoirs producing below bubble point. Oil rate is related to pressure drawdown empirically as following:

$$\mathbf{Q} = \mathbf{J}(\mathbf{Pi}^2 - \mathbf{Pwf}^2)^{\mathbf{n}} \qquad \text{Eq. 7}$$

Where:

J = productivity coefficient

n = empirically determined exponent, 0.5 < n < 1.0

Figure 12 shows flow rate and bottomhole flowing pressure versus time for a properly run flow after flow test. A log - log plot of flow rate, q vs different pressures should define a straight line with slope l/n. This plot can then be used to predict flow rate for any possible drawdown pressure as shown in Figure 13. At least four rates should be taken and each rate should continue until the well reaches a stabilized flowing pressure condition. This requirement for many wells means that long testing periods are required which often limits the usefulness of the flow after flow method range and apparently provides results comparable to the original flow after flow procedure. Transient Pressure Tests are probably the best way to determine true damage around a wellbore. The basic of pressure transient analysis assumes that the only well completed in a reservoir is shut in until a completely stable situation is reached. If this well is then put on production and pressure is reduced in the wellbore, a wave of reduced pressure gradually moves outward into the reservoir establishing a pressure gradient or sink toward the wellbore. With continued fluid withdrawals from the wellbore, the pressure wave moves further outward. Each point passed by the wave experiences a continuing pressure decline. At a particular time, the maximum distance the wave has travelled is called the drainage radius of the well. When the wave front reaches a closed boundary, pressure at each point within the boundary continues to decline, but at a more rapid rate. If the wave bent encounters a boundary which supplies fluid at a rate sufficient to maintain a constant pressure at the boundary pressure at any point within the drainage radius will continue to decline but a slower rate. With either the closed boundary or the constant pressure boundary the pressure gradient toward the well tends to stabilize after a sufficient time. Pressure level at a particular point may continue to decline. For the constant pressure boundary, a steady state will be approached where both pressure gradient and pressure level do not change with time. For the closed boundary a pseudo steady state condition is reached where pressure gradient is constant but pressure declines linearly with time at each point within the drainage radius. Changes in production rate or production from other wells will cause additional pressure wave movements which affect pressure decline and pressure gradients at every point within the drainage radius of the first well. The basis for transient pressure analysis is the observation of these pressure changes and the fluid withdrawal or injection rates which caused them, along with mathematical descriptions of the flow process, involving formation damage. Evaluation of damage where:

$$\Delta P = \frac{141.2\ Q.\beta.\mu.S}{kh} \qquad \text{Eq. 8}$$

Δp = pressure drop across skin, psi

p = formation volume factor, reservoir bbl/stb

p =viscosity, cp

s = skin factor, dimensionless

k = permeability, md

h = height, ft

The value of the skin factors can vary from about -5 for a successfully hydraulically fractured well to m for a completely plugged well. Negative values indicate stimulation and positive values indicate damage. One problem with the concept of skin effect is that the numerical value of the skin (S) does not directly show the degree of damage. Using the Homer plot, as shown in Figure 14, the skin factor can be calculated. Flow efficiency (or condition ratio) describes the wells actual flow capacity as a fraction of its capacity with no damage, as following:

$$FE = \frac{J\ actual}{J\ ideal} = \frac{P - Pwf - \Delta Ps}{P - Pwf} \qquad \text{Eq. 14}$$

Damage ratio is the inverse of flow efficiency. Skin factor, flow efficiency or damage ratio can be determined from most transient pressure techniques. Wells completed with only a part of the producing zone open because of ineffective perforating. This example is based upon the assumption that there is uniform fluid entry along the horizontal well length. Calculate the pressure drops in the skin zones in vertical and 1000 f long horizontal wells. The well tests show a skin factor of +1 for the vertical as well as the horizontal well. The following additional reservoir properties are given:

h =25ft,

kv=kh=l0md

qh = 2500 bbl/d

qv = l000 bbl/d

oil formation factor =1.06 rb/stb

viscosity=0.8 cp

Using equations (3) 6 and (4) 7 the results show 479 psi pressure drop in the vertical well and 30 psi in the horizontal well.

9. DAMAGE PREVENTION

By paying close attention to all aspects of well work formation damage can be kept to a minimum. Also, by keeping good records of materials and practices used, identification and detection of problems can be more accurate. Improved knowledge of the problems will result in improved efficiency in remedial treatments. The first opportunity we have to damage the formation is during the drilling of the well. This damage can be caused by invasion of the mud filtrate and particulate material in the mud. The particulate material comes from that added for weighting, for fluid loss control and other additives as well as drill solids. Earlier work reported the investigations of the effects of drilling mud particulate invasion on rock permeability can be summarized as follows:

a) Invasion and formation damage occur with all muds

b) The depth of invasion and level of impairment can be controlled, to a certain degree, by designing the mud to include bridging material

c) The effectiveness of the bridging material in reducing invasion is a function of the concentration and particle size of the material and of the pore sizes of the formation rock

d) Damage is most likely to occur in higher permeability formations; most muds contain sufficient quantities of particles, including cuttings, in the size range required to bridge lower permeability rocks

Where invasion occurs, back-flushing does not always remove the impairment high permeability mud cake. Even though the particulate solids may be too large to enter the pores they may not be sufficiently sealing to give a low permeability. It is thought a particle size distribution, some small and some large particles are needed to give a low permeability wall cake high degree of overbalanced pressure. When possible, this should be kept to a minimum. However, as has been discussed, a certain mud weight will be required to maintain hole integrity and well control. This may be especially true with horizontal wells. The underbalanced drilling primarily for horizontal wells can greatly reduce formation damage with advantages being reduced fluid loss and improved productivity (less damage). As always, concerns for well control should take priority when selecting mud weight.

Extended formation-drilling fluid contact time can obviously cause more fluid invasion and, therefore, damage. Again, this would be of concern primarily in horizontal wells. It can be an issue in non - horizontal wells when leaving an upper zone open while drilling deeper. So, when long exposure times are expected extra care should be taken to assure the mud is compatible with the formation or at least the damage zone is shallow.

Since most horizontal wells are not cased and perforated, damage prevention is of utmost importance. This can be best accomplished by providing an easily removed filter cake that has prevented excessive filtrate loss. The idea is to design drill-in fluids for controlling formation damage and mud clean-up in horizontal wells hole will assist in removing any filter cake that has collected. As mentioned above and in the previous chapters the potential formation damage problem with the actual cementing of the casing is the filtrate loss from the cement and preflush fluids to sensitive formations and formation fluids. A preflush or spacer fluid is designed to separate the drilling fluid and the cement slurry. The spacer must be compatible with both fluids while controlling both formation pressures and lost circulation zones. These fluids are usually high fluid loss and or abrasive fluids. This high fluid loss is expected to occur into the permeable zones and can cause damage. Spacer fluids are designed to thin the muds or to displace some of the mud and filter cake. Fresh water, brine, chemically treated water, emulsions or oil is used as spacer fluids. The base fluid chosen will depend upon the drilling fluid in the hole, the cement being used and the formation. Fresh water may cause formation damage from water blocks, emulsions and the hydration and or migration of clays and silts in the formation /127/. Brines may or may not be compatible with the cement slurry. Calcium chloride brines are not compatible with neat cement slurry. The brines generally cause less formation damage; however, they may form emulsions and can be incompatible with drilling fluids. A chemical spacer can prevent these problems if properly selected /128/. Oil based muds will normally require an oil-based spacer. This helps in protecting formation clays. Many oil based muds leave the casing and hole in an oil wet condition. For good cement bonding and hydrocarbon production it is preferred to have the system water wet. If possible, a water wetting surfactant should always be added to the spacer fluids. Mud displacement prior to cementing is usually conducted in turbulent flow. Where low viscosity fluids are used they are circulated in turbulent flow and contain mud thinners and dispersants that penetrate the mud and reduce its viscosity. Often the fluid will contain low concentrations of abrasive solids to scour the hole in order to help remove mud cake. The solids are especially useful when turbulent flow conditions cannot be obtained. Where higher viscosity fluids are used the objective is to remove the mud by viscous displacement. The higher the viscosity the more difficult it is to obtain turbulent flow reservoir permeability. This is especially true in overbalanced perforating where the fluid being used is not compatible with the formation, for example, where the drilling mud is the fluid in the hole when perforating. There is considerable overbalanced perforating now being conducted but it is with selected fluids and extreme overbalanced conditions. These concepts will be discussed more in the chapter on damage removal. Clean, non-damaging fluids should be used during perforating. Common perforating fluids are hydrochloric acid, organic acids, non-acid water-based fluids (e g., 2% KCl water) and oils. The concern with hydrochloric acid is its corrosiveness, particularly at high temperatures. This would require high concentrations of corrosion inhibitors which will oil wet sandstones and be damaging. The more common acids used for perforating fluids are formic and acetic acids. These are less corrosive and usually adequate for reaction with the perforation tunnel and the formation to prevent formation damage. Oil based perforating fluids can contain water wetting surfactants such that they will leave the formation water wet and may be a good choice when perforating very dirty sands. The API has a large data base available from standard tests that indicates the amount of permeability reduction caused by various perforating guns and charges. The amount of formation damage or permeability reduction from gun debris is difficult to estimate. Powdered metal charge liners were introduced to reduce damage from liner debris. The damage is finite and should be considered along with other factors. The perforation compacted zone is caused by the perforating charge firing and penetrating the casing, cement sheath and formation. The material in the path of the jet is broken up and pushed out of the way. Thus, the formation around the perforation tunnel is crushed and some reports show a permeability reduction of as much as 90%. Formation damage can occur during subsequent well completion work, following perforating as compacted zone is caused by the perforating charge firing and penetrating the casing, cement sheath and formation. The material in the path of the jet is broken up and pushed out of the way. Thus, the formation around the perforation tunnel is crushed and some reports show a permeability reduction of as much as 90%. Formation damage can occur during subsequent well completion work, following perforating as well as during workovers. Workover operations after perforating may also require non-damaging fluids. These workovers may require the killing of the well, so, well under specific test conditions. Absolute rated cartridge filters will achieve a sharp cutoff at their rated size. Most of the particles larger than their rating will be removed. The absolute filters will usually become plugged with particulate material much faster than nominally rated filters. Filter cartridge performance can be compared by using a Beta ratio. Beta ratio is defined as:

$$\beta = \frac{\textit{Influent Particle Count Greater Than Specified Size}}{\textit{Effluent Particle Count Greater Than Specified Size}}$$

Beta ratio is converted to Removal Efficiency as:

$$RE\% = 100\ \frac{[\beta - 1]}{\beta}$$

Beta Ratio	Removal Efficiency %
10	90.0
100	99.0
1,000	99.90 (Minimum Recommended)
5,000	99.98
10,000	99.99

When the fluid cleaning system where fluid returns from the well are taken over a shale shaker, through a particle settling tank overflowing into a residence tank, then pumped through the filtration section, consisting of two banks each containing two filter pots. The first and second banks contain nominal filter cartridges with 50-micron ratings and the second contains 2-micron absolute rated cartridges. The pots in each bank are manifolded such that one can be taken out of service to change filters while fluid continues to flow through the other for completion, stimulation and workover fluids to minimize formation damage. Also, where polymers are being used filtration may be necessary to remove insoluble polymer materials. Where the initial completion will include acidizing or hydraulic fracturing the formation damage from sources presented earlier are less important /88/. A major concern with these types of completions is the damage which may occur on subsequent well work, i e, loss of fluids which are dirty or otherwise incompatible with the formation. When pumping fluids down tubular for whatever purpose we should always be concerned about the material in the tubulars and it going to the formation creating damage. Of concern is scale from produced water, iron from mill scale and corrosion, pipe dope, paint, etc. By cleaning the pipe prior to pumping acids and other fluids to the formation some of this type of damage can be prevented.

Initial completions with horizontal wells in unconsolidated formations often includes a sand exclusion device (screen) without gravel packing In these cases the formation sand is allowed to collapse around the screen and sand control is accomplished with the screen. A potential problem with this completion is the filter cake material, formation fines, drill solids, etc plugging of the screens. The removal of the filter cake and other solids prior to allowing production through the screen is important to prevent screen plugging. Quality control in well work is important to prevent damage. It is not uncommon for formations to be further damaged when stimulation treatments are attempted. This can be caused by poor quality control at the well site as well as from the wrong chemicals being used for the formation being treated /106, 107/.

10. DAMAGE REMOVAL

Where formation damage has occurred and its effect is such that it should be removed, there are a number of different approaches, not all utilizing acid. One of these approaches is to bypass the damage during perforating. The industry standard for years has been to under balance perforate, i e., the pressure in the wellbore is less than the reservoir pressure. Here the objective is to let the formation pressure blow the perforating debris out of the perforation tunnels into the wellbore and to reduce the negative effects of the compacted zone around the perforation tunnel. The debris from the jet perforating gun is usually carbon, lead and copper. In some areas this approach has provided good result. However, there have been several reports of poor perforating efficiency using under - balanced perforating. One method which has been used to overcome this poor efficiency and perhaps some formation damage has been to use explosives (actually gas generators) to increase the efficiency of the perforations and to possibly bypass damage. A common commercial tool for this practice is the Stress Frac tool provided. Here the wellbore pressure reaches pressures much greater than the formation pressure and the fracturing pressures of the formation, usually 12 - 14 psi/ft, 27 - 316 KPdm. When the perforating guns are fired fractures are created around each perforation, perforating debris is blown out and away from the perforation. During this process frac sand or other proppant is placed in the small fractures when perforating sandstones. Acids can be used when perforating carbonates to develop flow capacity. Thus, the near wellbore area can be stimulated (or at least wellbore damage bypassed) on initial completion. So, it is possible on initial completion of cased holes that no other damage removal methods will be required. This concept may also allow evaluation of the reservoir and the determination of what, if any, additional stimulation will be required near wellbore damage such as a filter cake, the best approach may be to wash the filter cake off and dissolve, loosen any particulate material which was used during the drilling phase, using a fluid best suited for the mud cake and additives employed. For open hole completions, this washing could be done through the drill string on the way out of the hole leaving a completion fluid on the formation until the production string is run in the well. Of course, loss of the completion fluid to the formation during this time may be unacceptable and either viscosity and or particulate material may be necessary in the completion fluid to control the loss. For this reason, it may be preferred to run the production string and then wash the formation using coiled tubing to remove this near wellbore damage in an open hole completion. For cased hole completions, this washing may be taken care of with preflushes ahead of the cement. If, after the initial completion, the perforations have been plugged with mud or some other kill fluid anon- acid fluid may be of value in removing the damage, however, we would normally think about using acid on damaged perforations. The use of non-acid fluids normally would only apply if the damage was very near wellbore. The types of non-acid materials available for damage removal are numerous but can be classified as shown below:

a) Dispersants/Flocculants for Bentonite muds

Examples:

SAPP - Sodium Acid Pyro Phosphate Polymers

b) Oxidizers for Organic Materials

Examples

Bleach - Sodium Hypochlorite

SP - Sodium Persulfate

Cases they give preferred etching patterns in fracture acidizing of carbonates. Additionally, some iron control is provided. The normal organic acids used are formic, acetic and citric acids with formic and acetic being used at normally 10% by volume and citric as high as 100 lbs 11,000 gallons, 12 kg/m3 for iron control. The HCl in these blends can be from 5 to 28% with 7 to 15% being used commonly for near wellbore damage removal.

1. Organic Acids. The organic acids such as formic and acetic are advantageous at times due to their incomplete ionization, i e; they are weak acids. This 'retardation' provides a less corrosive environment while still having approximately 75% of the spending power of hydrochloric acid:

$$HCOOH \Leftrightarrow HCOO^{-}\ H^{+}$$

Formic Acid - 10% Formic ≅ 8% HCl

$$CH_3COOH \Leftrightarrow CH_3COO^{-}\ H^{+}$$

Acetic Acid - 10% Acetic ≅ 7 1/2 % HCl

These acids will have basically the same effect on carbonates and clays as HCl but the reaction rates are much slower.

2. HCl - HF Blends. Commonly referred to as HF Acid or Mud Acid. These are for sandstone formations only and should not be pumped above the fracturing pressures. The HF (Hydrofluoric Acid) will dissolve silicates, i e., clays, fine sand and shale particles and sand grains. However, the reaction on laboratory tests and some fields examples indicates an early damage in clay containing formations can occur when using HF acid. This early damage is related to the concentration of HF used, the amount of clay and fine silicate in the formation and the volume of acid injected In

general, the more clay a formation contains the lower should be the concentration of HF in the acid' /128/. When treating long intervals without proper zone isolation this early damage may cause self-diversion of the HF acid. This diversion may prevent sufficient volume of HF being injected into an interval. The end result may be increased damage rather than damage removal and stimulation. Extreme caution should be taken when using HF acid in wells with very little zone isolation. This would certainly be true in horizontal wells, particularly in open hole or liner only completions where zone isolation is difficult. Additives for use in acids are numerous but can be generally classified as shown below:

1. Surfactants. These are surface active agents which provide 3 basic functions:

a) Maintain water wet conditions

b) Lower surface/interfacial tension to assist in fluid recovery

c) Prevent, break or provide emulsions

Surfactants can be characterized as having a water-soluble head (hydrophilic) and an oil soluble tail (lipophilic). They are further defined as water soluble or oil soluble based upon the relative length of the oil soluble tail. Surfactants are further defined by the charge at the water-soluble end. This charge relates to the wetting characteristics on various types of formations

2. Fluid Loss Control Additives-Diverting Agents. These are important in the fluid loss control of a fluid which may be put on the formation and can be helpful in distributing acid throughout a zone being stimulated. The common materials available are:

a) Calcium Carbonate. This material is soluble in hydrochloric and the inorganic acids, formic and acetic. It obviously would not be used in an acid treatment. It is available in various particle sizes. The grade particle size range selected should be based upon the expected pore size distribution and laboratory testing. The advantage of calcium carbonate is that it is readily soluble in acid that comes in contact with it. Thus, it is one of the easiest of all fluid loss additives to remove after a well kill procedure or having been used for fluid loss control for drill-in/completion fluids in horizontal wells

b) Oil Soluble Resins. This material would be used in oil reservoirs or in situations where a solvent would be placed on the resin to remove it. It could be used in water-based fluids or acids. It is available in various particle sizes and melting point ranges. Care should be taken to make sure the resin is uniformly dispersed in the treating fluid

c) Benzoic Acid. Benzoic acid is slowly soluble in water or oil and will sublime in gas. It is available in finely divided particles and as flakes. For matrix fluid loss control, the finer material should provide the best control. As a word of caution, there have been reports of benzoic acid being washed out of wells 10 years after treatments where benzoic acid was used as the fluid loss/diverting agent material. Benzoic acid flakes may tend to agglomerate and pack together giving poor fluid loss control and difficulty in removing. The studies referred to here relate to treatments for perforation diversion using benzoic acid flakes. These materials are less affected by HCl than starch but most can be removed by oxidizers.
D. Wax Beads. These beads or buttons are malleable and will deform to form a seal on perforations or natural fractures. They are oil soluble and melt at specific temperatures.

E. Foam. Foam as a diverting fluid has shown good success in horizontal wells, in gravel packed wells and other applications. It is suitable in long intervals where particulate materials may not have application as well as in low pressure reservoir situations. Foams are normally 65 - 85 % liquid with the gas phase being nitrogen. The foam is stabilized with surfactants and often a gelling agent /161, 162, 163/. The surfactant can be any of the classes mentioned above. The above materials, except benzoic acid and the oil soluble resins, and wax can be mixed in either oil or water-based treatments fluids. Benzoic acid and the oil soluble resin and wax can be mixed in water or acid. Those which shouldn't be mixed in acid are the calcium carbonates and starches. Viscosity can be used for fluid loss control and to help in placement of the various fluid loss additives listed above. The common viscosifiers for this purpose are:

F. Hydroxyethylcellulose (HEC). This viscosifier can vary considerably based upon the degree of derivation. With the proper HEC and proper mixing and perhaps filtering, this polymer can be basically non-damaging. Acid is a breaker for this polymer as are oxidizers and some enzymes temperatures, but there is some control in viscosity reduction rate based upon temperature and acid concentration. The service companies providing these polymers can provide break data in acids. Using oxidizers to break these viscosifiers is also dependent upon oxidizer type, concentration, pH and temperature of the viscosified fluid. Enzyme breakers are more sensitive to temperature and fluid pH. Many are slower to react in high pH fluids and become much less effective in fluids with temperatures above 140° F (60° C). However, there have been many advances in enzyme chemistry for use in the oil industry and for many applications the above limitations are not valid.

3. Diverting Agents. These are very important when there is insufficient zone isolation by initial completion and/or mechanical means to make sure each zone is properly treated. The difference between diverting agents and fluid loss additives, as discussed here, is that diverting agents are for diverting treatments from one interval to another while fluid loss additives are for matrix control but could also be used for zone diversion. The same basic materials as the fluid loss additives listed above would be used for diverting agents with only a sizing difference, i e., larger size particles for diverting. Additionally, the viscosifiers would probably be utilized at higher polymer loadings and possibly crosslinked when used to place particulate diverting agents.

4. Corrosion Inhibitors. These are necessary to prevent excessive corrosion of the tubulars while doing an acid job. The type and amount will depend upon the acid type and concentration, temperature and contact time. Where matrix acidizing treatments are carried out at relatively low injection rates with large volumes the contact time can be very long. This long contact time may require chloride is added to the HF acid forming an aluminum fluoride complex. The decomposition of the AIF complex, under acidizing conditions forms HF acid at a relatively slow rate. Retardation is accomplished by controlling the availability of the HF for reaction. Dowell developed the use of fluoboric acid as a retarded acid system for sandstone. The fluoboric acid not only dissolves clays but also immobilizes clays and fines that are contacted but not dissolved. Fluoboric acid (HI3F4) hydrolyzes to generate HF acid and hydroxy fluoboric acid as follows:

Although the hydrolysis proceeds rapidly, the equilibrium allows only about 5 % of the available HF acid to exist at any one time. This equilibrium causes the reaction on the formation to proceed slowly for the same reason that the reaction of acetic acid is slower than that of HCl. The slower reaction rate allows the fluobotic acid to penetrate a greater distance into the formation before spending. After the available HF has spent, the remaining hydroxyfluoboric acid (HI3F3OH) slowly reacts with the remaining clays 'fusing' them and other fines to each other and to the sand grains. In addition to these methods of using HF acid to reach farther into the formation, it may also be possible to utilize a low concentration of HF, 1 112 % HF - 7 1/2 % HCl followed by a clay stabilizer such as a cationic organic polymer

5. Iron Control Additives. Iron can be brought into solution during an acid treatment from several sources. Both (ferrous) and (ferric) iron can be brought into solution by the reaction with various materials in the matrix such as hematite, magnetite, siderite, pyrite, iron stains, clays (chlorite and mixed layer), with connate water containing ferric or ferrous iron and with millscale and iron oxide in the pipe. Of these two forms of iron, the state is more difficult to acid (SWIC) which contains an iron chelating agent and sulphide control agent. The sulphide control agent helps prevent precipitation of elemental sulphur by preferentially reacting with H2S in the presence of iron. The iron chelating agent then chelates the remaining iron, holding it in solution. Thus, both sulphur and iron are maintained in a solubilised state in the treating acid

6. Mutual Solvents. Mutual solvents are often used in acidizing treatments for several reasons. The corrosion inhibitors used in acidizing are cationic in nature and will oil wet sandstones. One role of the mutual solvent is to prevent the adsorption of corrosion inhibitors on sandstones and leave them water wet. Additionally, where anionic surfactants are being used in the acidizing, they will tend to react with the corrosion inhibitors. Mutual solvents will help keep both of these materials in solution so they will function as they should. It is usually recommended for use in the overflush fluids unless field experience in the area suggests otherwise. If mutual solvents are used at temperatures above approximately 250° F, 121° C, additional corrosion inhibitor may be needed to control corrosion. Prior to any acid treatment, the condition of the treating pipe should be considered, whether the treatment will be down casing or tubing. Either of these may contain iron scales in any of several different forms, paraffin, pipe dope, scales, etc. It is usually advisable to clean the treating pipe with acid and a solvent prior to acidizing. A wellbore cleanout treatment will normally consist of circulating a quantity of organic solvent followed by HCl

acid through the tubing and up the annulus before setting the packer or reverse circulating down the annulus and up tubing when cleaning a casing string. The recommended organic solvent is usually xylene and the HCl concentration is usually 15%. The minimum amount of HCl is 250 gallons, 0 946 m3 with the usual amount penetrate the perforations and area of the formation with the highest permeability resulting in channels being dissolved. These channels are referred to as wormholes and because of the rapid reaction rate only a few may be developed unless steps are taken to divert the acid to other, lower permeability (or greater damage) zones. Without this diversion only a few of the perforations may be treated and those treated may not be treated adequately to bypass or remove all the damage. For this reason, fluid loss control and diversion in matrix acidizing of carbonates, particularly limestone's, is one of the most important aspects of getting a good damage removal treatment. Dolomites do not react with HCl as fast as limestone's and in some cases the acid should be heated to get the proper acid spending. This is usually when the dolomites are shallow, low temperature or the temperature of the acid is very cold. Several North American case histories have been reported where acid treatment success in dolomites could be correlated with the time of the year. The results have been better in the summer time when the acid temperature was warmer than those treatments conducted in the winter months. These were cases where low temperature, shallow depth dolomites were being treated. Even in these cases it is still important to consider fluid loss control and diversion of the acid to make sure the entire completion interval is treated. Some points to consider when conducting matrix acidizing treatments is the pressure during the treatment and the change in pressure when diverting agents are being used and which zones or intervals are being treated. Too often, when we pump acid into carbonate intervals we assume the entire interval is being treated. However, this is rarely the case unless fluid loss and diverting agents are being used. Fluid loss additives would be expected to stop the growth of wormholes and loss of acid to the matrix while diverting agents would be expected to divert acid from one set of perforations to another. The use of fluid loss additives should be considered when the acid reaction rate is fast, i e., when the formation is primarily CaCo3, limestone, and at elevated temperatures. The fluid loss additive should slow the growth of wormholes and give a more uniform distribution of the acid. The type and concentration of additive would

Psi/ft + 3000 psi/6000ft = 0.933 psi/ft. The 0.433 psi/ft is the hydrostatic pressure gradient for fresh water. With this as the frac gradient assumed to apply for the well to be treated, the injection rate and pressures for the matrix acidizing treatment would have to stay below this pressure. If the zone to be treated is at 6500 feet, the maximum bottom hole treating, BHTP, would be 6064 psi (0 933 psi/ft 6500 ft) With a full column of

15% HCl, the maximum surface pressure would be 3039 psi (15% HCl has a hydrostatic gradient of 0 465 psi/ft). The rate at which the acid treatment could be pumped would be controlled by this maximum pressure, the permeability of the interval under its current damaged condition, the length of the interval, number of perforations opened, friction pressure from the fluids being pumped down the tubular, etc. At matrix rates, usually less than 5 bpm, 0.8 m3/min, fiction pressures are usually negligible unless treatment is down small ID coiled tubing.

The advantages of matrix acidizing carbonates is that the damage is usually removed and there is less chance of communicating with unwanted production, i e., water or gas as could happen with fracture acidizing.

Carbonate Acidizing and Fracturing Acidizing

Many limestone and dolomite formations have low permeability's. Acid injection into these low permeability formations, even at moderate rates, may result in fracturing the formation. This type of treatment is widely used for well stimulation when acidizing carbonates and unwanted water is not thought to be a problem. The success of a fracture acidizing treatment depends upon the ability to create a conductive fracture having the appropriate height and extension out into the reservoir laboratory data. The prediction is difficult because it is dependent on such unknown factors as rock heterogeneity, fracture face roughness, acid displacement efficiency (viscous fingering), and other factors. Conductivity can be limited if the amount of acid reaction is insufficient, if rock strength is low (the fracture can collapse), or if the closure pressure is very high causing the fracture to close. Acid fracturing with HCl has application in carbonate reservoirs If hydrochloric acid solubility of the formation is greater than SO0?, then a good possibility exists that the well will respond to acid fracturing. Formations that are less than 80% soluble in HCl acid are not very good candidates for acid fracture treatments due to poor etching and the release of insoluble fines. Fracture acidizing in carbonates is done for three main reasons. The first and primary reason is due to low formation permeability where the well does not produce at economic rates without stimulation. A conductive fracture provides much more surface area for oil and gas to flow to the wellbore. The second reason for acid fracturing is to overcome damage that extends to a greater depth than can be removed with matrix acid and the third reason is to break down perforations in a sandstone reservoir and thus improve flow efficiency. Due to the difficulty in getting optimal results in fracture acidizing carbonates several improvements have been developed to help with the success. The two most important are viscous preflushes and density-controlled treatments. Viscous preflushes have been used for several years to improve upon fracture acidizing results. A viscous, water gel preflush will cool the formation face, assist in controlling fluid loss and protect a portion of the fracture face from the acid which follows. The end results from all this is more effective use of the acid. Control of fluid density during acid fracturing can be an extremely important aspect of an acid treatment. Fluid density differences exist during fracture acidizing. These density gradients affect where the live acid is placed in the fracture and, therefore, where unwanted gas production should have a preflush with density less than the acid his will cause the acid to under ride the preflush and migrate toward the bottom of the created fracture.

Acid movement toward the lower portion of the fracture helps to develop conductivity there, and not as much as in the upper portion of the fracture. Density control acid fracturing treatments on thick zones, such as 100 or 200 feet, 30 48 or 61 meters from top to bottom perforation, should have the acid placed toward the center of the fracture. Two preflushes can be used to control the acid movement toward the midsection of the fracture height. A higher density preflush should be pumped first. The light density preflush follows behind the first preflush. Acid with a density between the two preflushes will tend to stay toward the middle of the created fracture, without migrating up or down. Fracture growth during an acid fracturing treatment is a dynamic process. Preflush moves away from the wellbore creating the fracture and some of it is lost to the matrix or natural fractures as fluid leakoff. Acid reacts with carbonates along the fracture wall and is increased in density due to reaction products, such as calcium chloride. As this acid reacts, it is increasing permeability and accelerating fluid loss into the fluid wall. Without controlling this loss, the depth of penetration of live acid will be very small. The service company representatives who did the original work on density control acidizing recommend that the preflush for these treatments be replenished every 10-15 minutes throughout a treatment. Evidence that fluid density does influence acid placement was found in a West Texas well. A whole core was taken from a Level land field well when it was deepened 30 feet. The core contained a fracture which was etched with acid from a large volume, high-rate acid treatment conducted prior to coring the well. The conclusion drawn from this core evaluation was that fluid density did influence acid placement and that the acid did preferentially go down damage which is beyond the filter cake. Depending upon the type and the depth of the damage it is possible to use an acid other than HF. Hydrochloric acid and organic acids will tend to dehydrate and flocculate clays, particularly the smectitic clays which make up bentonite. If damage is due to very near wellbore swelling clays (filter cake) then it may not be necessary to dissolve the clay and a hydrochloric or organic acid could be used /131/. Additionally, due to the HT acid spending on calcium carbonate and calcium magnesium carbonate resulting in a calcium fluoride precipitate (CaFl), there is a general rule of thumb that HF acid should not be used on formations containing more than 15 - 20% calcium carbonate or calcium magnesium carbonate. For damage that has been caused by mud filtrate invasion (disrupting clays and other silicates deeper in the formation) and damage from fines, clays, etc., migrating to the wellbore area during production, HF acid may be the preferred treatment. When this is the case, the necessity of zone isolation takes on an added significance. As was discussed early in this chapter, it is thought to be important to keep HF acid and its reaction product away from calcium, potassium and sodium. A preflush ahead of the HF acid is required to push connate water away from the HF reaction area and to dissolve any carbonate in this area. When carbonate is present, the usual preflush fluid is hydrochloric acid at low concentrations, 5 - 10%. When no carbonate is present, other fluids such as ammonium chloride, diesel oil, etc. may be used. Obviously, potassium, sodium or calcium chloride should not be used. The HF acid stage follows this preflush and is followed by an after 1 overflush. The objective of the overflush is to keep the HF moving and spending on untreated formation. Thus, each fluid must go into each interval being treated in the proper order, i e., preflush, HF acid, and overflush, at a pressure below fracturing pressure. As envisioned, it may be difficult to obtain proper zone isolation so each fluid goes into the zone being treated in the right order. The longer the exposed interval the more difficult it is to control an HF acid treatment. So, the completion method becomes very important to the success of an HF treatment. Obviously, successful HF acidizing in horizontal wells would be very difficult. In general HF acid in treatment than before and the profit from the well increased in a formation where the pore spaces are small and poorly interconnected, i.e., low permeability, the fluid trying to move to the wellbore will be restricted and resulting production rates will be low. By putting a high flow capacity fracture in the formation, the fluid can more easily move to the wellbore. This is a simple analogy but brings up one important point, the formations that respond the best, i.e. greatest production increase from hydraulic fracturing, are those that have low permeability. If a formation can produce high rates through high permeability, then a fracturing treatment is probably not going to be of much help except to bypass damage.

11. MUDS AND TREATMENT FLUIDS

Muds generally refer to all circulating mediums in the bore-hole that are controlled during the drilling process. Drill-muds can be subdivided into groups according to their composition. Figure 9 Enclosed shows the density grading in different drill-muds.

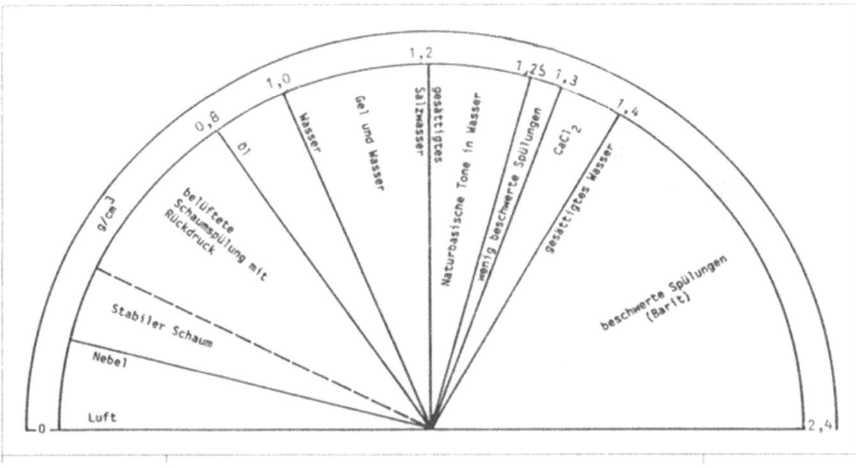

Figure 9: According to API /4, 5/ drill-muds are classified according to their composition

A drill-mud must fulfil the following functions /2, 4, 5, 7, 147/:

1. Complete cleaning of the floor of the borehole,

2. Transportation of the cuttings to the surface,
3. Keeping the drill particles in suspension in the case of a break in mud circulation,
4. Elimination of the drill particles on the surface
5. Control of the formation´s pressures,
6. Development of a thin and low permeable filter-cake,
7. Reducing of the formation damage,
8. Prevention of fall-outs,
9. Cooling and greasing of the drill equipments,
10. Prevention of corrosion,
11. No impairment of the implementability and assessment of geophysical borehole measurements and
12. Transfer of the hydraulic performance on to drill motors

Apart from this, it must be guaranteed that the mud:

1. does not require any large-scale completion measure,
2. does not endanger the health of the drill personal and
3. does not pose any potential technical environmental danger

According to their solid substance content, drill-muds are subdivided into "solid substances poor" and "solid substances rich".

Drill-muds with poor solid substance contents which consist mainly of a liquid phase and polymers cause high loss of fluid in high permeable rocks. In muds rich in solid substances, the solid substances make up its weight and serve in the development of a filter-cake /132/. These muds normally consist of the liquid phase, weight-medium, bridging materials and filter-reducing supplements. Weight-building-materials regulate the drill-muds and can lead to the building of bridges, which can result in the reduction of fluid-loss in the Formation. Solid substances to be used in formation-areas must be soluble in oil, water, or acids. Heavy and insoluble materials like barite and Ferrous-oxide are not suitable for use in formation-areas /4/.

Treatment Fluids

Frac-Fluids

The Frac-treatment of a Formation should improve its productivity. This function is fulfilled through the hydraulic building of cracks in the formation, which enlarge the drainage space and break through the permeability barriers. The functions of the Frac-fluids are the transmission of the pressure required to break the rock and the transportation of the of the supporting-mediums into the already created cracks, to keep them open for a long time /14, 76, 86, 149 - 156/.

Frac-liquids are subdivided into the following groups:

1. Water based liquids
2. Oil based liquids
3. Emulsion-, and Foam-, as well as
4. Acid-based liquids

Frac-fluids are to ensure that with small volumes, the desired crack-lengths are achieved through low loss of filtrates. A sufficient viscosity is of special interest to the transportation and distribution of the supporting-mediums in the crack-system, as well as the creation of the desired crack-width. A likely high density with similar injection-density at the head of the borehole should make a high Frac-pressure in the formation possible. Apart from this, the

Frac-fluids should be stable in terms of temperature and shearing and should not cause any formation damage.

To adjust the Frac-liquids to the desired characteristics, the following additives come into use /25, 76-78, 86, 92, 148-150, 152-155, 157, 158, 161-163/:

- pH buffer
- Filtrate reducer,
- Stabilisers to prevent fall-outs,
- Gel stabiliser,
- Gel breaker,
- Network connector,
- Bactericides,
- Clay inhibitors
- Substances active to the boarder surfaces and
- Friction reducer

Filtrate-reducing additives have the function of binding free water and thereby guarantee a minimisation of the formation damage caused by the Filtrate.

In order to prevent or at least delay the precipitation of calcium-carbonates and calcium-sulphates, Polyphosphates can be used.

Gel-stabilisers have the function of preventing the thermal decomposition of Polysaccharides by temperatures above 100°C.

Gel-breakers serve the destruction of the networks in treatment-liquids after a definite period of influence, in order to be able to further the formation fluid back again.

Network-connectors cause an overlapping between the polymer-thread molecules, which is like the building of a net and lead to an increment of the viscosity.

Since the Solvents of native Polymers represent a good fertile soil for bacteria, water-based Frac-fluids are displaced with bactericides.

For the stabilisation of the clay, 1-3% calcium-chloride or Amine is added to the treatment fluid.

Water-based Frac-fluids

More than 70% of all Frac-treatments are currently being carried out with water-based liquids. Beside the comparatively low costs and the simple handling the water-based liquids distinguished themselves through a high productiveness. /29, 30, 59, 76, 86, 148, 150, 154, 157, 160/.

In water-based Frac-fluids, the native Polymer Guar-gum is frequently applied as Gel-medium Figure 10.

A)

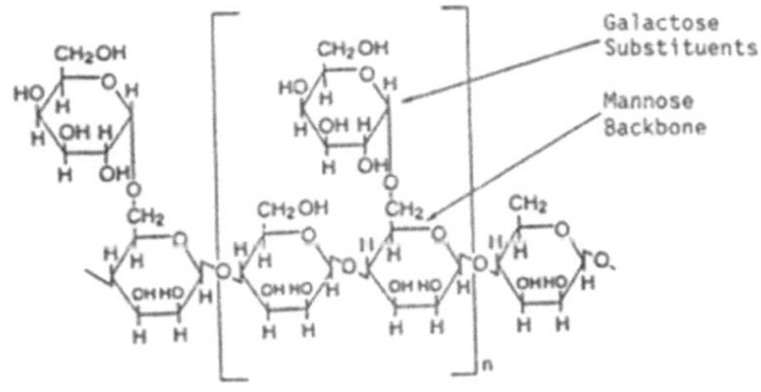

Figure 10: Guar-gum Structure

After dissolution in water, the macromolecules hinder themselves through overlapping in their movement-ability, which leads to a rise in viscosity.

Guar-gum is very productive and processes a high salt-stability; but it is sensitive to bacteria attacks. Guar-gum contains 6-10% of insoluble components that can cause the blockage of the flow channels of the formation. Through the deposition of side chains on the polymers, derivatives like Hydroxipropylguar (HPG) are produced that show only 2-4% residues. HPG distinguished itself through high thermal stability /59, 76, 83-85, 150/.

A further often used Polymer is Hydroxyethylcellulose (HEC). The advantage of this product lies in the production of a very pure, residue-free polymer-solution. Enclosed document 6B shows the structure-form of a HEC-monomer /25/.

Xanthan, the expensive biopolymer produced by the micro-organisms represents the latest stand of the development Figure 11.

B)

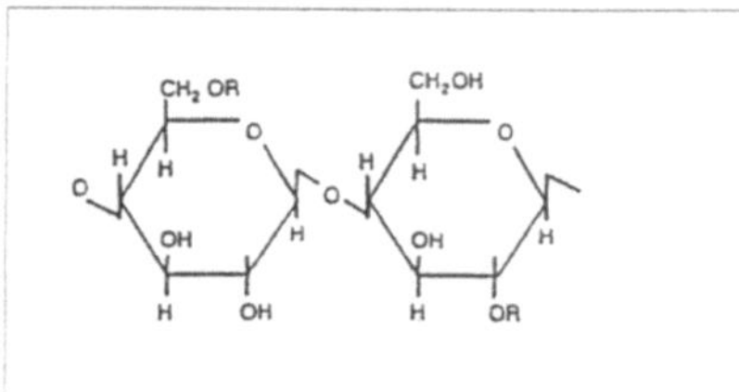

B) Strukturformel von Hydroxyethylcellulose (HEC) /25/

C)

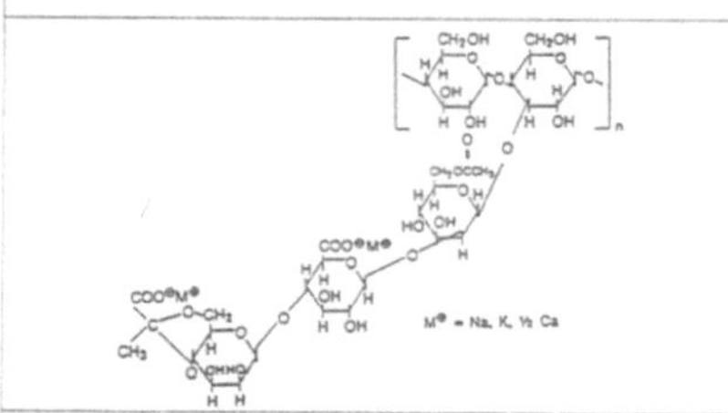

Figure 11: The Biopolymer Xanthan Structure

In order to achieve a high starting-viscosity with less polymer-concentration, the frac-fluids are treated with network-connectors (bridging-ions). Borate, Titan-, Aluminium-, Zircon-, and Antimon-complex connections are used as “cross-linkers”. Apart from this, the addition of these network-connectors guarantees a temperature-stability of the frac-liquid of up to 200°C. Because a high viscosity is desired in the area of the boreholes and the cracks, the connection-reaction is delayed.

The system polymer-connector-delayed will be harmonised alongside in line with the treatment conditions /25, 76, 84, 86, 92, 158, 161-163/

Oil based Frac-fluids

Oil based Frac-liquids are only introduced, when a very strong formation-damage through water-based formation-fluid is expected. To achieve the required viscosity, a jellying of the oils with Aluminium-phosphate-tester is carried out. Differences in viscosities could be adjusted depending on the quantity-components of aluminium-complex and phosphate-testers.

Emulsions and Foams as Frac-fluids

Water-sensitive formations can necessitate the employment of multi-phase systems as frac-fluids, service firms offer foams and emulsions for this /25,76,111,150,164/.

Foams are stable dispersions whose exterior liquid phase is made up of oil or water. Up to 95% of the foam volume in the interior-phase is made up of gas. Nitrogen preserves itself at brings less danger of corrosion alongside. Tensiles are employed to ensure the stability of the 2-phases system.

Polymers and network-connectors can come into use for further stabilisation. The advantages of foams can be seen in the good transportation characteristics, the very low loss of filter, the low formation damage and the fast removal of the frac medium after the stimulation /25,76,149,164/.

Emulsions preserved themselves as frac-fluids because of the lower formation damage. The so-called poly-emulsions with oil making up their interior phase are most frequently used; the oil content is 67%. The stability of the emulsion is brought about on one side through the jellying of the exterior, watery phase with polymers /25,72,76,120,137,149,153,157/.

Acid-based Frac-fluids

The following acid based frac-fluids make it possible to control the loss of filtrate successfully:

a) Mixtures from acids, polymers and oil-resins as well as

b) synthetic polymers that retain their acid sources and their

other characteristics.

Acid-oil emulsions (60-90% acid phase) can also be employed. A mixture a nitrogen-organic-base and non-ionic tensides are used as emulgators.

The selection of frac-fluids depends mainly on the types and contents of deposits. Especially the deposits-parameters temperature, pressure, moisture-capacity, proportion of saturation have influence on the choice of fluids.

The loss of filtrates should be minimised because the filtrate pressed into the formation is no longer available for the expansion of the split-system; apart from this, a formation-damage is caused through the filtrate.

The relation of the generated split-volume to the whole volume of the formation-fluid can be used in judging the filtration relation of the frac-liquid. To obtain a desired split-length, a lower volume of frac-liquid is required for a better filtration relation (lower loss of filtrate) /59, 92, 49,152/.

The viscosity of the frac-fluid is decisive for the transportation of the support-medium into the split. High viscosities of the frac-fluids lead to bigger split-widths, while the split-length is reduced. To ensure lower loss of pressure and good pumping characteristics, the viscosity in the area with high shear stress (pumping, mixer and perforation openings) must be low /14,76,86,92,149-151,153,155,156,157,160/.

In Newton's liquid, the sinking velocity of the particles of the solid substance according to Stoke`s law can be calculated as follows:

$$V = \frac{d^2 . g}{18\,\eta}\,(ps - pf) \qquad \text{Eq. 15}$$

g	=	Earth's acceleration
d	=	diameter of the particles of the solid substance
η	=	dynamic viscosity of the liquid
ρs	=	density of the solid substances
ρf	=	density of the fluid

Gravel Formation-liquids

Low viscose liquids like the Newton's fluids (water, salt-solutions or oil) also come into use as gravel formation liquids, the high viscose fluids like polymer solutions (HEC, HPG and Xanthan Gum) are given preference because of certain advantages /82,164,165/.

Hydroxyethylcellulose (HEC) is the most widely spread polymer, that is used as viscosity-raising additive for gravel formation-liquids.

The flow-relation of gravel formation-liquids is important in view of the bearing capacity, especially at the entry (over cross-over valve) into the ring space, where the flow-velocity decline and as such the bearing capacity for the filter-medium could also be reduced. Particularly at this stage, a high viscosity as well as a sufficient jell-strength of the formation liquid will be required for the transportation of the gravel /82,165,166,167/.

The degree of hydration of XC, HPG and HEC solutions depend on salt content, temperature, extent of shearing during mixing and the pH-value of the basis solution.

The choice of the required jell-breaker for the polymer is determined through the type of components of the formation liquid (polymers, salts etc.) as well as the temperature level.

Acids

Through acids, permeability-barriers that arise through the influx of solid substances and caused through rock-formation are disposed off. Principally, there are three possibilities to achieve this aim /14, 90,168-178/:

1. Addition of acidic solid substances to the drill-muds,

2. Through the dissolution procedures of the corresponding parts

of the matrix and decomposition of diagenetically newly developing minerals in the pore spaces.

3. to create additional splits that serve as flow-routes for the pore-contents in the rock.

The first two possibilities mentioned can be realised through acidification and the last through frac-operations. As shown in

Figure 12, the effect of an acidification in the production of a high transparency zone around the bore hole can be seen. The improvement that is thereby to be expected can be described through the following relations:

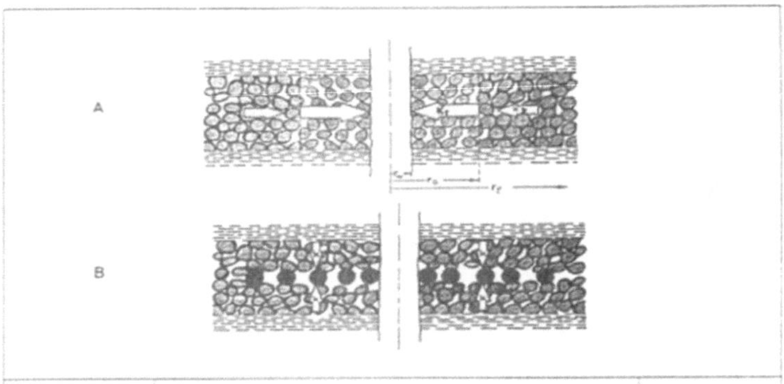

Figure 12: The effect of Acidification

$$\frac{Q}{Qo} = \frac{\frac{k1}{k2}\ln\left(\frac{re}{rw}\right)}{\ln\left(\frac{ro}{rw}\right)+\frac{k1}{k2}\ln\left(\frac{re}{rw}\right)} \qquad \text{Eq. 16}$$

Q	=	Influx-rate after treatment
Qo	=	Influx-rate before treatment
k1	=	Transparency of the rock in the acidified zone
k2	=	Transparency of the rock in the untreated state
rw	=	Radius of the bore-hole
ro	=	Radius of the acidified zone
re	=	Entrance radius

Through an acidification, an increment of influx according to equation 10, only with carbonic connections, but not without further ado in the clay-minerals is to be expected.

Through the employment of HCl, Alkali, Soil-alkali and aluminium ions can be attacked, while SiO2 remains or is transformed into amorphous form respectively; further obstructions of the rock can also occur alongside. Also, to remove SiO2 one also frequently liquid acid that form water soluble or gas-forming siliceous-hydroxide-fluorine-acids. The use of liquid acid has a disadvantage, in that it can combine with calcium ions to form calcium-fluoride because HF does not react immediately with SiO2. The calcium-fluoride is practically insoluble. If the liquid acid reacts with the calcium-ions afterwards, precipitations occur; this could lead to blockage of the rock. The supporting-medium and the acids intrude into the splits that arise in the case of acid-frac and prevent the full closure of the splits. Figure 2B.

The cleaning of a borehole is understood as chemical or mechanical removal of the borehole coating (filter cake). The removal of the filter cake is important for a better adhesion of the cement to the mountains. In case chalk rinsing is used the filter cake can be removed with hydrochloric acid /175, 177, 179, 180/:

$$CaCO_3 + 2HCl \leftrightarrow CaCl_2 + H_2O + \uparrow CO_2 \qquad \text{Eq. 9}$$

12. CONSTITUENTS OF THE DRILLING MUD AND TREATMENT FLUIDS

Clays

Clay minerals originate from the weathering of magmatic rocks. The original minerals are feldspars [(CaO) (K2O) Al2O3 6SiO2], ferromagnetic minerals as well as hornblende [(Ca, Na2)2(Mg, Fe, Al)5(Al, Si)8 O22(OH, F)21]. Bentonite arises from the weathering of volcanic ashes. More than 50% of the clay minerals are Illites.

The most important types of Clay are /4/:

a) Montmorillonites and Illite-Montmorillonites,
b) Chlorides and Chloride-Montmorillonites as well as
c) Kaolinites, Septochlorides

The clays can basically be subdivided as follows:

a) Two-layer-clays
b) Three-layer-clays
c) Clays with double-chain-structure

Clay minerals show two structural units in the crystal lattice /2, 4, 5, 99, 181-186/. One block consists of SiO4 components coordinated in a tetrahedral way linked to a second layer by oxygen atoms. This second layer shows an octahedral coordination and contains Al-, Mg-, or Fe-atoms in the gravital centre of the components.

If the layers of a three-layer-clay are marked with A, B and A from top to bottom, the A-layers consists of SiO4-tetrahedrons while the B-layer consists of Al-octahedrons. The composition of B- and A-layers to a three-layer-clay is illustrated in Figure 13. As you can see from this figure the link between the B- and A-layers in a three-layer-clay is made by the oxygen-atoms through the ends of single tetrahedrons; these are therefore characterized as bridge-oxygens.

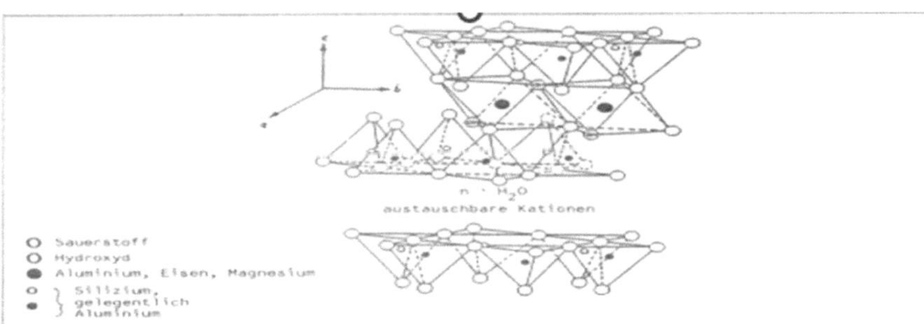

Figure 13: Build-up of Montmorillonit/7/

Structure

According to its structure, Montmorillonite belongs to the three-layer-clays consisting of an inner aluminate-layer (Al-O-OH-layer) and two outer silicate-layers /186,187/. Aluminate-layer and silicate-layers are linked to each other by oxygen-atoms, the so-called bridge-oxygens and form an elementary layer.

Figure 14 schematically shows an elementary layer which as structure-former does not have any importance for rinsing due to its charge equalization.

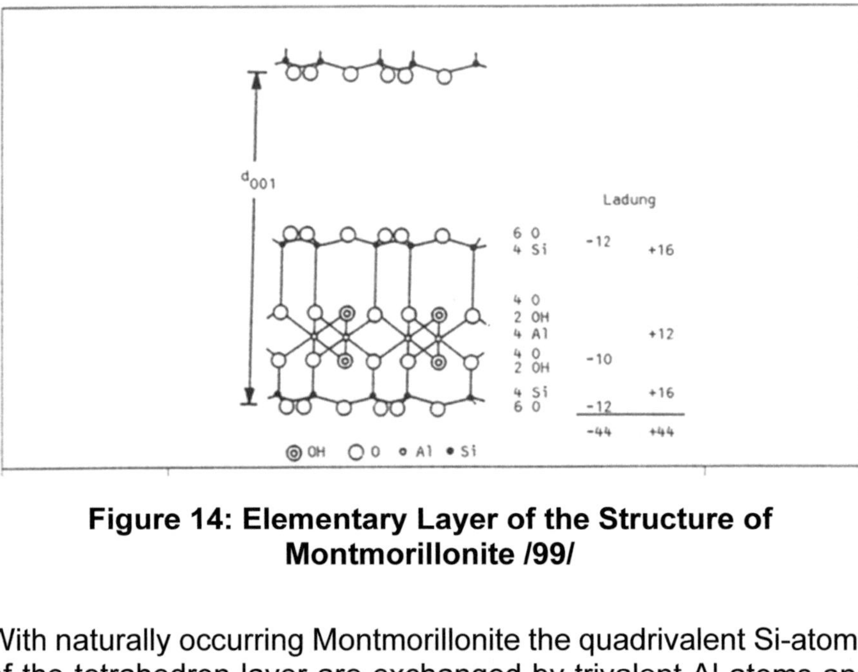

Figure 14: Elementary Layer of the Structure of Montmorillonite /99/

With naturally occurring Montmorillonite the quadrivalent Si-atoms of the tetrahedron-layer are exchanged by trivalent Al-atoms and the trivalent Al-atoms of the octahedron-layer by bi-valent Mg- or Fe-atoms in an irregular sequence. By this isomorphic substitution a negative excess charge on the basal plane of the elementary layers occurs due to the lack of positive charge, which is equalized by absorption of cations like Na+, Li+, K+, Ca++, Ba++, Mg++, H+/4, 5, 184/.

The adhering and diffuse ions of the double layer are characterized as opposed ions; these can be exchanged by others. Concentration, validity and exchangeability are decisive to the exchange of ions. Na- and Ca-ions are due to their size assembled to the surface, while smaller Li-atoms can find room in the octahedral gaps. By their oxygen affinity, H-atoms transform Si-O- -groups into Si-OH-groups. The mostly assembled cations determine the physical and chemical properties of the clays; these are therefore named after the most assembled cations (Na, Ca-, Montmorillonite, etc.).

Through break of the elementary layer parallel to the C-axe primary bonds are released in the tetrahedron- and octahedron layers, which on one side of the break have the form ≡Si+, =Al+ and on the other side ≡Si-O- =Al-O-.

The examinations carried out by VAN OLPHEN on aluminium brine showed that Me-O-bonding's are in the acid area neutralized through absorption of H+, while Me+-bonding's continue to exist and cause a positive charge of the brine. He transferred these results on clays and assumed that the break edges of the clays must have positive charges.

The examinations of RUEHRWEIN und WARD /188/, PACKTER /189/, SCHOTT /190/ and SCHWARTZ /191/ on the influence of polyelectrolytes on Montmorillonites infer positively charged break edges and confirm the assumption of VAN OLPHEN. RUEHRWEIN and WARD proved the assembling of polycations to the negative basal planes of the elementary layers of Montmorillonite through increasing the distance of the elementary layers with the help of x-ray diffraction; SCHWARTZ showed that no increasing of the c-distance of the elementary layers of Montmorillonite occurs through addition of polyanions.

Since this was provably an adsorption of polyanions their assembling to the positively charged break edges of the clay had to be assumed.

The results of the rheologic examinations of the Dehydrill HF-rinsing, particularly after an electrolyte- and/or temperature load, confirm the belonging of the DHT to the Bentonite-(Montmorillonite)-group /147/.

Through ascertainment of the IR-transmission spectrum, which among others proves the existence of silicates and hydroxyl-groups (OH) /146/, the same characteristic values could be proved as already stated for hectorite by Van OLPHEN /184/ as well as BECK and BRUNTON /192/.

The indication of the lamellar structure (d0001-distance of appr. 1.1 nm) of DHT by the RDA-method proves it's belonging to the flaky-lamellar structured Montmorillonites; this structure is also confirmed by REM shots /147/.

The examined DHT lamellas show sizes up to 120 µm basing on a tile-shaped flaking of many single components. The maximum size of clay is appr. 2 µm /147, 193/.

The exchangeability of cations of DHT was determined 38 meq/100 g according to API-norm RP 13 B (methylene-blue method /194/). This value corresponds to the exchangeability of cations of hectorite /195, 196/.

According to literature /197/ Hectorite is a tri-octahedron three-layer-clay with the sum formula:

$$[(Li_{0,66}\,Mg_{5,34})\,Si_8\,O_{20}\,(OH)_4]^{0,66(-)}\,.\,0,66\,Na^+ \qquad \text{Eq. 10}$$

During synthesis the tetrahedron-layers stay seized with silicon, while in the octahedron's magnesium is partly replaced with lithium. We know, however, hectorites with which the Si-atoms of the tetrahedron-layer were partly replaced with Al-atoms /198/.

The concentration of K+, Li+, and Ca+ with DHT is, compared to the Na+-content, very low. However, the seizing-density of assembled cations is lower in DHT, resulting in a higher richness.

The seizing of the octahedron-gaps of the dehydrill HT by Li+ and Mg 2+ can also serve as explanation of the higher richness /199/.

As a summary, we can conclude that Dehydrill HT (hectorite), which is synthesized at 180°C, belongs to the group of saphonites. The most important elements participating in its structure are Si, Mg, Na and Li /200/:

$$MgO \,.\, a\,MA \,.\, b\,Al_2O_3 \,.\, c\,SiO_2 \,.\, n\,H_2O \qquad \text{Eq. 11}$$

M = Na^+ and/or Li^+ with a Na/Li-relation > 1

A = F, OH and/or 1/2 O_2

a = 0.1 to 0.6

b = 0 to 0.3

c = 1.2 to 1.7

n = 0 to 3.0

Hydratation

The process of swelling is made possible through mudding the clay in water. According to /2/ three phases are distinguished:

1. Directed assembling of water-dipole-layers to the basic planes of clay with little supply of water. This is called "dry-solid" consistency ("non-liquid water")

2. Additional absorption of water between the clay particles, which allows the clay to be plastically distorted

3. Supply of excessive amounts of water which lead to the formation of a clay suspension

With clay suspended in water the spreading density of the assembled cations and the concentration of electrolyte in the water are responsible for the cohesion of the elementary layers /129, 201-204/. With declining density of charge the bonding energy between the elementary layer's declines.

The declining bonding energy between the assembled cations and and the basic clay planes results in a concentrated penetration of water-dipoles into the gaps of the elementary layers. The assembled cations diffuse into the surrounding liquid because of thermal motion and osmotic pressure. They thus form a diffuse layer of ions around the elementary layer/the elementary layer-package Figure 15 (2, 99, 185). The water-dipoles then occupy the places having become vacant on the basic clay planes.

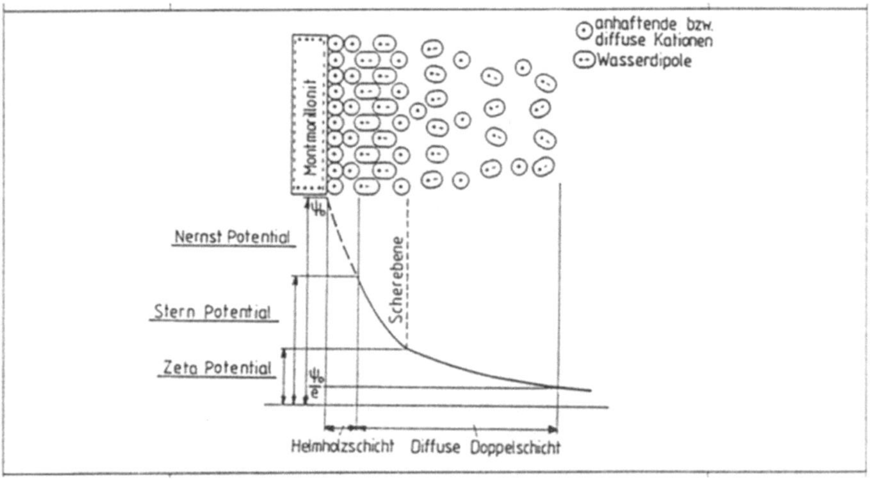

Figure 15: Diffuse layer of ions around the elementary layer/the elementary layer-package/147/

According to BRANDLY, GRIM and CLARK the thickness of a dipole-layer is 0.3 nm. With dry Montmorillonite, the c-gap of the elementary layers is 0.92 – 0.98 nm /99/. Through assembling of water-dipole-layers the d001-gap of the elementary layers of Montmorillonite can be increased up to 4 mm Figure 16.

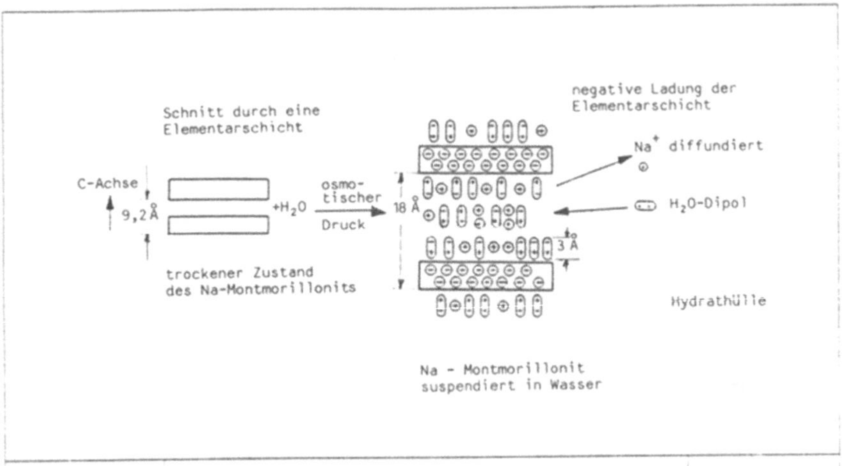

Figure 16: Assembling of water-dipole-layers/99/

Through hydratation (increasing of the d001-gap) the bonding energy of the elementary layers of the clay diminishes, so that they separate from each other and form independent particles (intercrystalline swelling).

An absorption of water by clay is also affected by surface-hydratation of the elementary layers, which is called "inter-particular swelling" /187/.

In case the particles are subject to heating the hydratation is accelerated. The swelling ability of the clay decreases rapidly depending on the concentration of electrolytes of the water and the valency of the ions /2, 17, 99, 187/.

Through the progressive development of the filter-cake, the value of W increases, this leads to a reducing frictional force between the external layers of the filter cake (compare equation 1). An equilibrium eventually emerges (see equation 2), such that the development of the filter-cake gets interrupted; a constant filtration-rate emerges afterwards.

$$R \leq \mu * A\,[\Delta P - \left(W \pm \frac{2\,.\sigma\,.\mathbf{Cos\theta}}{r}\right)] \qquad \text{Eq. 20}$$

$$Ks \geq R \qquad \text{Eq. 12}$$

R = the frictional force between the particles of the solid substances in the cake formation

μ = Number of frictions between the components of the solid substances

A = Surface of filtration

ΔP = Differential pressure

W = Pressure-loss in the pore-channels of the cakes and/or of the infiltration zone

σ = Surface border tension between the filtrate and walls of the pore.

r = Radius of capillarity

Φ = Angle of Moisturisation.

Ks= the shear-force exerted by the circulating mud of the filter-cake

A temperature and/or an electrolyte strain of the applied clay-mud affects the degree of hydration of the clay and polymer negatively. A dehydration leads to a change in the filtration-characteristics of the drill-mud through an excessive supply of free water as well as a change in the number of friction and the nature of deposition of the clay particle in the cake-formations. These changes result in an intensified formation-damage.

Polymers

In order to regulate the losses of water, to stabilise the clay in case of electrolyte- and temperature loadings and to control the flowing behaviour, polymers are added to the clay muds /2, 4, 5, 7, 14, 25-27, 34, 147, 181-183, 205-209/.

Polymers are short- or long-chained molecules with non-ionic or anion-active groups like R-COO-, -SO 3. They are subdivided according to their origin (figure 2).

Native and semi-synthetic polymers (encl. 6A-D) belong to the group of "polysaccharides". The combination of single monomers through "acetal combinations", to who's dissolving a specific work of only 335 kJ/mol is necessary, is characteristic of this group. /205, 209/.

With synthetic polymers (encl. 6E), the polymerisation is mainly affected through C-C-bonding's, to whose dissolving an incomparably higher specific work is needed.

Structure

According to MOORE and HUMMEL /209/, a polymer is understood as substance whose molecules according to their formula are a multiple of a basic constituent, the monomer (An), which allows assuming different values. Binary and ternary systems as well as copolymers are those containing two, three or more different basic constituents. Depending on their distribution alongside the chain, they are called altering [(Am Bn) x; m and n are differently low numbers], or block-polymers [(AmBn) x; m and n are different numbers here].

Concerning their structure, the polymers used in the mud technology differ in two key data determining their properties:

the rate of polymerisation indicates the average number of monomers being unified in a filamentary polymer molecule. It is responsible for the properties of a polymer concerning rheology.

the rate of substitution is understood as average number of active anions per monomer of the filamentary molecule. It hints at the stabilising effect of the polymer on clay.

Hydration

In the statistically most probable form, polymers in dry state are present as "balls" /7, 147, 205 - 209/. If you put polymers into a polar solvent, the following factors will become effective as described in 2.2.3.1.2:

a) decline of concentration,
b) Brown's molecular motion and
c) electrostatic forces of attraction

Caused by the decline of concentration, the Na+-ions diffuse into the surrounding liquid and lead to the formation of a cation-cloud around the filamentary molecule. Water-dipoles are taken up into the occurred gaps, which causes the hydration of the molecules. The diffusion of the Na+-ions into the surrounding liquid in irregular order leads to an activation of the electric charge of the R-COO-, which cause an electrostatic repulsion alongside the chain; therefore, an increased stretching of the filamentary molecule occurs, so that this can turn from a ball through intermediate stages into a stretched form (encl. 13). If the polymers are in a flow, Brown's molecular motion is superimposed by a shearing force caused by the flow; the molecules are additionally set in rotation /2/. These cause a stretching of the molecule in the direction of the flow and a compression perpendicular to it.

On one hand, this effect leads to a quicker hydration of the polymers; on the other hand, the reduction of the viscosity of the polymer-solvents dependant on the shearing incline is also a consequence of it.

Further Additives

Borates, complex titan-, aluminium-, circone- and antimony-compounds are used as "cross-linker". Solution and swelling of these additives is as for the other polymers. Peroxydi-sulfates, which through forming of radicals from a temperature of 50 °C on decompose the polymer-chains, are mainly used as gel-decomposites in water-basic treatment-liquids besides enzymes. Therefore, this salt is added only in very low concentrations with higher deposit temperatures. The molecule chains of the aluminiumphosphate-ester-compounds are decomposed in oil-basic fluids with the help of weak acids or bases.

13. STABILITY OF DRILL MUDS AND TREATMENT FLUIDS

Theories on Stability

The term "stability" of water basic drill muds means their ability of resisting changes within the system. Generally, clay muds are no stabile systems. Coagulations of the disperse phase (area-to area-addition of clay-particles), which are synonymous with their dehydration possibly leading to the muds becoming unusable, take place more or less rapidly.

The velocity of the coagulation besides the factor time depends on type and concentration of electrolyte, temperature, as well as on the intensity of the shearing force /204/. These parameters can lead to overcoming, reduction or even elimination of "hindrances" counteracting the coagulation of the clay particles. The particles can then approach to an extent that the forces of attraction between them prevail /99, 201, 204/.

In order to better understand these processes, the theories on stability of colloidal systems shall be examined closer in the following:

Energy of Interaction

The theory of "energy of interaction" originally comes from HAMAKER /210/; it was later developed by OVERBEEK, VERWEY, DERJAGUIN, and LANDAU /195/.

When two clay particles approach each other parallelly in a suspension, their energies of repulsion as well as of attraction become effective; their resultant is responsible for the stability of the system Figure 17.

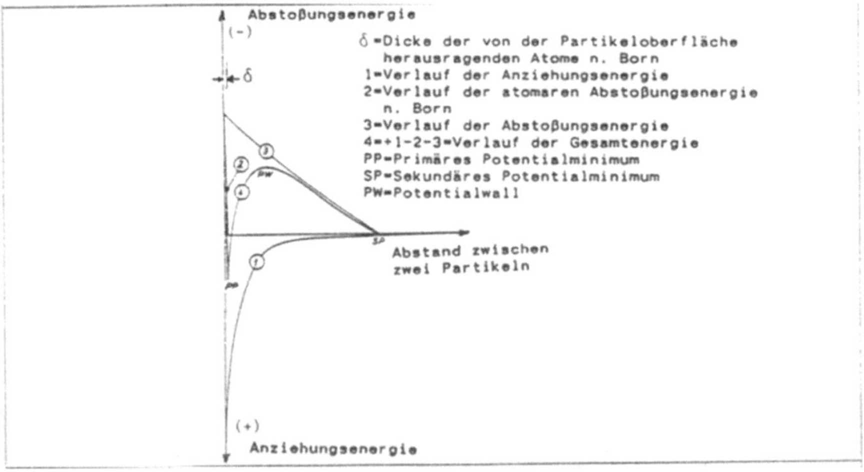

Figure 17: The Energy of Two Clay Particles

The energies effective in the zone nearby are:

a) the London-van der Waal's energy of attraction
b) the electrostatic energy of repulsion and
c) the atomic energy of repulsion according to Born

The London-van der Waal's energy of attraction is effective in the outermost area of the nearby-zone and responsible for the sticking of the coagulated clay particles. Changing dipole-moments of an atom effect the polarisation of another atom, so that these attract each other mutually. Since this attraction takes place between neutral particles, it is a quantum mechanical phenomenon, which according to HAMAKER /210/ solely depends on the distance between the two clay particles. Electrolyte and temperature do not have any influence on its level /2/.

$$\mathbf{VA = -A/(48\,\pi\, d^2)} \qquad \textbf{Eq. 13}$$

$$\mathbf{A = (3/4)\,(\pi\, q\, \alpha)^2\, h\, f} \qquad \textbf{Eq. 23}$$

A = Hamaker's constant

VA = potential of attraction

d = distance between clay particles

q = number of atoms per cm^3

α = Polarisability

h = Plank's effective quantum

f = basic frequency of the atoms

The electrostatic repulsion is based on the interference of electric double-layers of two clay particles approaching each other. In doing so, the distribution of the ions changes, whereby the potential/the free energy of the system increases /2, 211/.

According to OVERBEEK /211/, the effective repulsion results as follows:

$$VR1 = \left(\frac{64\,n\,k\,T}{x}\right).\left(y^2.\,e^{-x.d}\right) \qquad \text{Eq. 14}$$

V R1= potential of repulsion

n= number of ions per volume-unit

k= Boltzmann-constant (R/NL)

d= distance

R= universal gas-constant

NL= Lohschmidt's number: 2,687 * 1025 (l/mol)

Equation 21 shows, that the energy of repulsion depends on temperature as well as on valency and concentration of the electrolyte /2, 99/.

Born's atomic energy of repulsion VR2 is based on the unevenness of the basic area of the clay particles. It prevents complete indenting of the coagulated particles.

Thus, the resulting total energy of the interaction is as follows:

$$Vges = VA - (VR1 + VR2) \qquad \text{Eq. 15}$$

The curve of the total energy Figure 17 shows three extremes:

- primary minimum of potential (pp)
- potential wall (pw) and
- secondary minimum of potential (sp).

As indicated already, the coagulation of two clay particles approaching each other parallelly is only possible if the potential wall can be surmounted and the particles come to rest at (pp). This process is irreversible. Dependant on concentration of the electrolyte and valency of the ions, the amount of the potential wall can diminish. In case the potential wall reaches the field of attraction, an area-to-area-addition can take place at any collision of clay particles of any kinetic energy. With rising temperature, the energy of repulsion increases; the potential wall becomes bigger. Depending on the temperature, the kinetic energy of the particles /2/ increases in a way that at a critical temperature the potential wall is surmounted; an area-to-area-addition is the consequence.

$$\mathrm{Ekin} = \frac{3}{2\mathrm{kT}} \qquad \text{Eq. 16}$$

If the kinetic energy of the clay particles is not sufficient to surmount the potential wall, the approaching particle can come to a standstill in the (sp). This quasi-stabile state is seen as a cause for the formation of gel /2/.

Kinetics of Coagulation

The theory of "Kinetics of Coagulation" originates from SMOLUCHOWSKI /212, 213/ concerning monodisperse systems and from Müller /198, 214/ concerning polydisperse systems.

In it, the stability of a mud is judged according to the number of collisions leading to an area-to-area-addition per time unit /99, 215/.

The velocity of coagulation is mainly determined by /216/:

- Brown's molecular motion and

- the effect of power between the particles.

Through Brown's molecular motion, a zigzag-like motion of the clay particles is affected, in which these get into the sphere of activity of other clay particles. Depending on the level of energy of repulsion being effective between the particles, this approach can either lead to coagulation or not.

If, due to an influence of electrolyte, the energy of repulsion is reduced as far as the energy of attraction prevails in the nearby zone, each approach of two particles leads to coagulation. This case is called "rapid coagulation". Otherwise, a slow coagulation is present.

The number of particles reaching the sphere of activity of a fixed particle in a certain time can be calculated with the help of the first and second Fick's law /99, 216/.

$$\mathbf{n = 4\,\pi\,D\,RA\,n^\circ\,t} \qquad \text{Eq. 17}$$

$$\mathbf{D = k\,T/(6\,\pi\,\gamma\,r)} \qquad \text{Eq. 18}$$

n° = starting number of particles

D = diffusion constant

T = absolute temperature

k = Boltzmann-constant

r = radius of particle

γ = viscosity of liquid phase

t = time of coagulation

RA = radius of effectivity of energy of attraction

According to STAHLBERG /99/ and GHOFRANI /2/, the following relation between the number of coagulated particles after the time t and the starting number of the dispersed particles is given:

$$n = n_o e^{-4 \pi D R_A n_o t} \quad \text{(Eq. 29)}$$

$$n = n_o e^{-f t} \quad \text{(Eq. 30)}$$

$$f = 4 \pi D R_A n_o \quad \text{(Eq. 31)}$$

The time during which the number of particles originally existing in the suspension is reduced by a half through coagulation is called time of coagulation (tk).

The time of coagulation of the fast coagulation (t ks) is stated as follows by SMOCHULOWSKI:

$$\mathbf{tks = 3 \gamma r/(2 k T R_A n_o)} \quad \text{(Eq. 32)}$$

During "slow coagulations", not every approach of two clay particles leads to a coagulation, since London-van der Waal's energy of attraction as well as Coulomb's energy of attraction are effective.

The time of coagulation of the slow coagulation (t_{k1}) is longer than the coagulation time of fast coagulation by the reciprocal of the factor α.

$$\mathbf{Tk1 = 1/4 \pi D R_A n_o \alpha = 3 \gamma r/(2 k T R_A n_o \alpha)} \quad \text{(Eq. 33)}$$

$$n = n_o \, e^{-f \alpha t} \qquad \text{(Eq. 19)}$$

α = 1 is the case of "fast coagulation". With α = 0 the mud is completely stabile; 0 > α > 1 is the case of "slow coagulation" /2/.

Stability under Difficult Drilling Conditions

The drilling muds are exposed to high loads of temperature, electrolyte, and shearing depending. The stability of drilling muds is more or less reduced by these loads.

Temperature Load

As described already, an increase of viscosity takes place at first with unprotected clay suspensions after a decline of viscosity of the liquid phase depending on the temperature /2, 4, 5, 7, 99, 181-183, 191/. This increases up to the maximum viscosity has its cause in an intensified edge-to-area-addition of the clay particles. With further increase in temperature, the water-dipoles of the hydrate-sheath of the clay particles will be sheared off more intensively; therefore, the viscosity decreases while the rate of filtrate is increasing. After exceeding a critical temperature load a coagulation of the clay particles can happen; this process is irreversible /33/.

With a clay suspension protected by polymers, a temperature load leads to an increased tendency to vibration of the filamentary polymer molecules, which leads to a higher rate of stretching of these molecules. Since the kinetic energy of the system increases through heat input, the bond energy between the water dipoles (D-D) of 5-20 kJ/mol can be surmounted at first /7, 209/; the water-dipoles bonded in the hydrate sheath of the polymers dissolve, so that the viscosity of the system decreases. Through further increase in temperature, the bond energy between the water-dipoles and the charge carriers along the polymer chain, which is 40-130 kJ/mol, can be exceeded /7, 99, 191, 209/. Thus, it comes to an intensified separation of water from the hydrate sheath. The consequence is a further decrease in viscosity of the mud besides a further increase in free water.

In case the load temperature is less than the critical temperature of the system, the separation of water is reversible. In case the temperature exceeds the critical value of 40 kJ/mol, a thermal and/or oxidative decomposition of the polymer takes place (depolymerisation) /2, 4, 5, 7, 99, 191, 205/. The decreased viscosity and increased rate of filtrates having taken place this way are irreversible /4, 5, 7, 99, 191/.

Shearing Force

Through pure shearing, i.e. through addition of mechanic energy, dipole-dipole-, dipole-ion-, and ion-ion-bonds can be decomposed alongside the polymer-chain; thus, similar effects as mentioned above have to be taken into consideration in case of a shearing force.

Electrolyte Load

Electrolyte valencies and concentrations affect the stability of the clay suspensions through change of the hydration rate of the particles. The iso-electric point, i.e. the state in which the concentration of a cation on the surface of the particles is as high as in the liquid phase, plays an important part in that.

The addition of electrolyte to a drilling mud can, depending on the amount, either result in exceeding the iso-electric point or not /2, 99, 191/.

Above the iso-electric point, two cases can be distinguished:

1. If the cations of the clay and those of the electrolyte are different, the cations on the surface of the clay can be replaced with those of the electrolyte (for example transformation of Na- to Ca-bentonite with addition of Ca-salts).

2. In case the exchangeable cations of the clay are the same cations of the electrolyte, an intensified compression of the cation-cloud will take place, resulting in an increased release of already bonded water and thus effecting a decrease of viscosity.

If the iso-electric point is not exceeded with addition of electrolyte, a decline of the concentration gradient will take place; cations having already diffused into the liquid phase will be pushed back. Water-dipoles bonded in the hydrate sheath of the clay particles and/or the filamentary polymer molecules are released gradually. The ζ-potential of the clay particles and/or the polymers decreases.

In order to stabilise clay suspensions against the above-mentioned influences, polymers can be added. Concerning this,

Filtrate of the sand core and

protective colloids

are distinguished.

Filtrate droppers are polymers being able to bond only the free water of the drill. Since these are not able to dissociate, filtrate droppers are used in order to control the filtration properties of the drill.

Conditional on the increase in viscosity of the system, however, a slight stabilisation of the clays (formation of a mechanical barrier) can be expected from the filtrate droppers.

Protective colloids are poly-electrolyte dissociating in water as charge carriers. The structure of their hydrate sheath is similar to that of clays. After their dissociation, the anion-active filamentary molecules are taken up by the positively charged breaking-edges of the clay particles. With addition of protective colloids to a clay suspension, two cases can be distinguished:

The concentration of protective colloids is low. Several clay particles have to be taken up by a filamentary molecule; the clay lamellas. This process is called "sensibilation".

Comparatively much protective colloid is added to a clay suspension. The positively charged breaking-edges of the clay particles are occupied with a corresponding number of filamentary polymer molecules. Only in the liquid phase can the surplus polymers take over the task of bonding water.

Correspondingly, the protective colloids stabilise the clay suspensions through mechanic (hydrate sheath) as well as electrostatic barriers (ζ-potential). Therefore, a higher kinetic energy of the particles is necessary to surmount the barrier, which is synonymous to a better state of stability.

14. FORMATION DAMAGE CALCULATIONS

The effect can be calculated using Darcy´s equation for the average permeability for the zone near the wellbore that

$$K_{avg} = \frac{k_s k_e \ln(r_e / r_w)}{k_s \ln(r_e / r_s) + k_e \ln(r_s / r_w)} \qquad \text{Eq. 34}$$

Or

$$\frac{J_s}{J_o} = \frac{k_s / k_o \log(r_e / r_w)}{\log(r_s / r_w) + k_s / k_o \log(r_e / r_s)} \quad \text{Eq. 35}$$

is damaged (or stimulated), permeability in the near wellbore area, damaged (skin) or stimulated (Ks), and the natural permeability of the formation. Average permeability through reservoir and near wellbore damage (Kavg). Permeability away from wellbore, in the undamaged or non stimulated area (Ke,Ko). Drainage radius for the reservoir feeding the wellbore (Re). Radius of the damaged or stimulated area around the wellbore (Rs). Radius of the wellbore (Rw). Production rate through the damaged or stimulated well (Js). Production rate without damage or stimulation (Jo).

$$Q = \frac{7.07kh(P_e - P_w)}{\mu \ln(r_e / r_w)} \quad \text{Eq. 36}$$

By using Darcy´s equation for radial flow we can determine, using Kavg from equation (1) that the production rate expected for a damaged well and what could be expected if the damage is removed or perhaps the permeability increased in the near wellbore area. Equation (2) allows for a comparison of the productivity ratio for a damaged or stimulated, near wellbore, situation to an undamaged well. Darcy´s equation for radial, steady state, liquid flow is giving where; production rate, bbls per day (Q); permeability, darcies (K); formation thickness (h), feet; **p**ressure in the reservoir, psia(Pe); **p**ressure in the wellbore, psia (Pw); **v**iscosity, centipoise (μ); **d**rainage radius for the well, feet (re); **w**ellbore radius, feet (Rw).

$$Q = \frac{7.07kh(P_e - P_w)}{\mu \ln(r_e / r_w)}$$

Eq. 37

As an example, using equation (1), assume a well is located in reservoir with a drainage radius, re, of 660 feet (201 meters) with a wellbore radius, rw, of 0.33 feet (0.1 meters), a permeability in the damaged zone, Ks, of 1 millidarcy and a permeability in the undamaged formation, Ke, of 100 millidarcies with the damaged zone radius, rs, being 2 feet (0.61 meters). This is depicted in with this damaged around the wellbore the average permeability, Kavg.

$$K_{avg} = \frac{1 * 100 * \ln(660 / 0.33)}{1 * \ln(660 / 2) + 100 * \ln(2 / 0.33)} = 4.1 \text{ md}$$

Eq. 38

Using Darcy´s equation for oil flowing which has a viscosity of 1.0 cp with a pressure drop, (Pe-Pw) of 100 psi the production rate, Q, would be calculated.

$$Q = \frac{7.07 * 0.0041 * 15 * 100}{1 * 7.6} = 5.7 \text{ bbl per day}$$ Eq. 39

If the well were undamaged or damage removed by acidizing the production rate would be another rate.

$$Q = \frac{7.07 * 0.1 * 15 * 100}{1 * 7.6} = 139.5 \text{ bbls per day}$$

Eq. 40

Also, if the damage around the wellbore was removed and the original, undamaged permeability was increased to 200 md (for the 2 feet radius) in the process the production rate could be calculated, and for the same draw down, i.e., not much production increase for doubling the permeability over the original in this 2 feet interval.

$$K_{avg} = \frac{200 * 100 * 7.6}{200 * 5.8 + 100 * 1.8} = 113 \text{ md}$$

Eq. 41

and for the same draw down

$$Q = \frac{7.07 * 0.113 * 15 * 100}{1 * 7.6} = 158 \text{ bbls per day}$$

Eq. 42

i.e., not much production increase for doubling the permeability over the original in this 2 feet interval. Using the Js/Jo ratio (equation 2) and the same values in the above example,

$$\frac{J_s}{J_o} = \frac{1/100 \ \log(660/0.33)}{\log(2/0.33) \ + 2/100 \ \log(660/2)} = 0.04$$

Eq. 43

And where Jo= 139.5 bbls per day, Js=5.6 bbls per day.

These examples all relate to actual formation damage often referred to as skin damage. When a skin value is determined, a postivie value indicates damage and a negative value indicates stimulation. The skin effect, S, was introduced by Van Everdingen nad refined by Hawkins and is defined as:

$$S = \frac{(k_e - k_s)}{k_s} \ln(r_s / r_w)$$

Eq. 44

This equation points out that the degree of permeability reduction is more significant than the depth (radius) of the damage.

These examples all relate to actual formation damage often referred to as skin damage. When a skin value is determined, a postivie value indicates damage and a negative value indicates stimulation. The skin effect, S, was introduced by Van Everdingen and refined by Hawkins Figure 18.

This equation points out that the degree of permeability reduction is more significant than the depth (radius) of the damage.

The total skin would include any flow impairment from mechanical, well configuration effects. This can be additional pressure drop, sometimes referred to as pseudo-skin effects, and can be caused by inadequate perforations, tubing I.D. too small, surface equipment, etc. Only attributing the total skin to damage inside the formation is a common mistake. The pseudo-skin, if present, must be subtracted from the total skin to estimate the true skin associated with real formation damage. Pseudo-skin is also referred to as rate a sensitive skin where pseudoskin exists, as the production rate is reduced, the pseudo-skin is reduced. Acidizing or other stimulation attempts would not be expected to reduce this type of damage. Another rate sensitive skin is that from turbulent flow caused by producing wells at high flow rates resulting in turbulent flow in the perforations and/or in the formation. Additionally, producing oil wells at a high drawdown can cause the pressure at the wellbore to drop below bubble point resulting in gas evolution, an additional phase (gas) and a reduced relative permeability (to oil) near the wellbore. Producing wet gas wells at high drawdown can also cause an additional phase (condensate) to come out of solution resulting in a relative permeability effect. To prevent permeability damage or to stimulate a well effectively, it is important to accurately

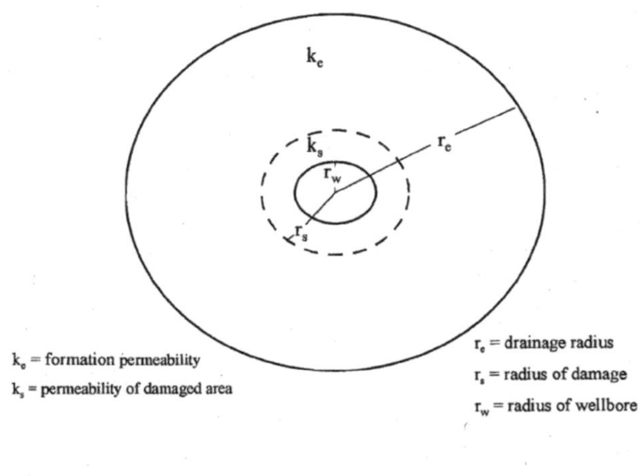

Figure 18: Damage area around the wellbore

In practice, the differential pressure leads to the invasion of the formation by the mud, solid substances of the mud and mud-filtrates, such that the original permeability becomes reduced /1-13, 15-28, 31-72/. Through this, there is an increased flow resistance against the flow of petroleum and gas respectively in the damaged zone.

This expressed itself as additional pressure drop in the pressure development within the reservoir, which is referred to as "skin-effect" /4, 9, 10,11/.

Muscat /12/ shows that the pressure drop in radial flow is directly proportional ln (re/rw). For the productivity of a sonde after the damage applies:

$$\frac{\mathbf{Qd}}{\mathbf{Q}} = \frac{\ln\left(\frac{\mathbf{re}}{\mathbf{rw}}\right)}{\left(\frac{\mathbf{ku}}{\mathbf{kr}}\right).\ \ln\left(\frac{\mathbf{rd}}{\mathbf{rw}}\right) + \ln\left(\frac{\mathbf{re}}{\mathbf{rd}}\right)} \qquad \text{Eq. 20}$$

Where:

Qd	=	Productivity after the damage
Q	=	Productivity before the damage
re	=	entry radius or the sonde
rw	=	radius of the damaged zone
rd	=	borehole radius
ku	=	permeability of the formation before the damage
kr	=	permeability of the formation after the damage

From this equation, it is obvious that the negative influencing of the productivity depends on the permeability course kr/ku and the radius of the damaged zone rw. The so-called Damage Ratio (DR) can be used to assess the damage:

$$DR = \left[1 - \frac{kr}{ku}\right] * 100\% \qquad \text{Eq. 21}$$

ku = original permeability

kr = rest-permeability

The borehole conditions, the compound of the mud as well as the characteristics of the horizons of the formation are to be considered as the influence parameters of the formation damage. The individual influence factors have already been listed under point 2.

15. TECHNICAL EXPLANATION OF THE EXPERIMENTS

The level of formation damage and the extent of the damaged zone depend on the properties of the fluids in use, the well conditions and the physical and chemical properties of the formation rocks or the substance contents of the pores. The level of the influence of the fluids on the formation damage depends upon:

- The proportion of the solid substances in the fluids
- grain form and grain size distribution of the solid substances
- mineralogical composition of the solid substances
- type and concentration of the polymers, as well as
- viscosity and density of the fluids

Well conditions that can influence the intensity of the damage are:

- formation temperature
- flow velocity of fluid
- differential pressure
- duration of contamination, as well as
- type and concentration of the electrolyte solutions that can get into drill mud and treatment fluids through kicks.
- Geology and dispositional environment and
- Reservoir properties

Physical and chemical properties of the formation rocks and substance contents of the pores are also relevant factors in connection with damage:

Formation rocks:

- permeability
- porosity
- capillarity
- size and geometry of the capillaries
- mineralogical composition and
- moisture capacity

Substance contents of the pores:

- type
- constituents
- viscosity
- flow characteristics, as well as
- pressure and temperature of the pore contents in situ

16. IDENTIFICATION OF THE FORMATION DAMAGES

In practice, formation-damages ("skin effect") are ascertained through injection. Also pumping experiments with subsequent evaluation of the corresponding pressure development as well as geophysical borehole measurements (see encl. doc. 2a) can give information about the extent of the influence of the damage. A further possibility is the comparison of neighbouring exploration drillings.

17. DAMAGE MEASURMENT AND CALCULATION

The extension and the level of the formation damage caused by solid substances and / or filtrate can be determined through permeability measurements before and after the damage.

Damage Ratio (DR)

The formation damaging effect of the applied liquids can be quantified through the "Damage Ratio".

For the calculation of the Damage Ratio (DR), it is necessary to experimentally determine the original permeability (ku) of the sample and its remaining-permeability (kr) after the damage (see equation 4).

Sectional Damage Ratio (SDR)

The penetration depth of the damage can be judged through the Sectional Damage Ratio (SDR). To this, the sample will be segmented after the damage, and the remaining-permeability (kri) of the individual segments will be measured. The measured values can be used to calculate the Sectional Ratio with the help of equation 8 :

$$SDR = \left(1 - \frac{kri}{ku}\right) * 100\% \qquad \text{Eq. 47}$$

Drilling Muds and Treatments Fluids

Beside drill-muds, this section will also deal with treatment liquids, gravel-formation liquids and acids.

18. OBJECTIVE OF THE EXPERIMENTAL WORK

Drilling muds and treatment fluids can reduce the productivity of the carrier horizons more or less through invasion of solids and/or release of filtrates. This phenomenon described as "formation damage" in the terminology can be caused physically, chemically and/or bacterially. It is the objective of this study to determine the level and the depth of penetration of the formation damage caused by selected drilling muds and treatment fluids; furthermore, the causes of the discovered damage shall be analysed.

In order to determine the level and the depth of penetration of the damage, the parameters "Damage Ratio (DR)" and "Sectional Damage Ratio (SDR)" shall be used. The possibilities of scanning electron microscopy (REM),

Thin grinding at the polarising microscope and of chemical analysis with the help of atomic absorption spectrum analysis shall be used in order to judge the causes of damage.

The following facts as influence factors of the formation damage shall particularly be paid attention to:

- differential pressure
- temperature
- velocity of circulation
- duration of contamination and/or treatment
- petrophysical and petrochemical properties of the carrier

rocks and

- composition and rheology of the drilling muds and treatment fluids

19. EXPERIMENTS PROGRAMM AND ANALYSIS

The programmes for the analysis to be carried out within the scope of the thesis and this book are described in enclosures 19 only in the book and others up to 23 as part of the program of experiments not showing in the book.

Encl. 19 describes the programme for analysing the influence of salinity and pH-value of the penetrating water on the level of permeability of the core specimen /133/. Therefore, non-damaged Bentheim- and/or Oberkirchen- sandstone cores are damaged with deionised water, tap water or KCl-solutions of different concentrations in each case. These core-specimens are then penetrated with air and/or tap water, deionised water, a 3 % of the weight of NaCl-solution, KCl-solutions of different concentrations or a 2 % of the weight of CaCl2-solution. The penetrating with air lasts 4 hours. The specimens are applied by 20 PV with fluids as penetrating media. Then, the specimens are tested for permeability, pH-value, damage ratio (DR) and the calcium concentration in the filtrate.

The examination of DHT-, bentonite-and polymer-muds) in view of the filtration behaviour is made with the experimental programme described in encl. 20. Drill muds of different compositions (encl. 19) are electrolyte-loaded with CaCl2 of different concentrations and then exposed to a temperature load of 90 °C for 24 hours. Velocity of circulation, differential pressure, circulation temperature, counterpressure and duration of the experiment as experimental conditions are varied at the circulation machine (see encl. 19). The Bentheim- and/or Oberkirchen-sandstone cores having been damaged in this way are then tested for damage ratio (DR) and sectional damage ratio (SDR). Scanning electron microscopic exposures (REM), thin grinding and atomic absorption spectrum analysis (AAS) follow.

The programme for examining frac-fluids and gravel formation fluids in view of their flowing behaviour and the formation damage is shown in encl. 21.

Frac-fluids and gravel formation fluids of different compositions, which have to be prepared according to special mixing-conditions (see encl. 21), are tested for their damaging effect on Bentheim- and/or Oberkirchen- sandstone with the help of the circulation machine. Differential pressure and circulation temperature as experimental conditions are varied. After damage, the specimen are tested for their DR and SDR. With certain core specimen, REM exposures follow. Before the experiment is carried out, the flow curves of the used fluids are registered with a rotation-viscometer (Fann, model 35 S). From this, the plastic viscosity and Bingham's yield point of these fluids can be calculated.

The programme for determining the level of change of damage degree of core specimen after re-extraction attempts is described in encl. 22.

Core specimen of Bentheim sandstone are damaged with a fluid of special composition in the circulation machine. While differential pressure, circulation temperature and duration of the experiment are constant for these experiments, the velocity of circulation is varied from 0 to 0,6 m/s. The damaged cores are tested for their permeability in a fluid-permeameter. Now, the cores will be taken out of the permeameter and once again inserted into the Hassel's cell after having been turned by 180°. In this phase, the measuring medium (1,5 % of the weight of KCl-solution) finds its way in through the end of the core, which does not contain any filter cake. Displacement pressures of 10 - 200 kPa are adjusted and according values are established.

In encl. 23, the programme for analysing the level of residual damage after an acid treatment of the cores of Bentheim- and

Oberkirchen- sandstone having been damaged with drill muds is described.

The damaged cores with which the DR was established are penetrated through in the Hassel's cell with 10 PV acids of different compositions and concentrations under a certain differential pressure. Following, the permeability values of the core and/or the core segments are determined in order to be able to calculate the DR- and/or SDR-values.

20. EXPERIMENTAL EQUIPMENT

Circulation Machine

The circulation machine (Encl. 24 A) having been further developed at the ITE is used for analysis of dynamic and static filtration as well as formation damage through drill muds and treatment-fluids.

Essentially, this machine is composed of:

a filtration cell and

a mud circulation

The filtration cell is a filtration autoclave working according to the principle of a Hassel's cell. Here too, the 5 - 25 cm-long core specimen (13) are enclosed by a rubber cuff (12), whose sheath pressure is generated by water over a manual pump (14) and read from a manometer. In order to not exceed a certain pressure, the machine is equipped with a safety valve (15). The filtration cell and the mud circulation are equipped with a sheath heating (11), with which temperatures up to 96° can be achieved. Oil, whose temperature is adjusted over an oil thermostat (5), serves as heat transfer medium. The oil is heated in a store container and conducted through filtration cell and mud circulation over heat exchangers.

With the help of a flow controller (10), the drift velocity of drill muds and treatment fluids can be varied through modification of the column/fissure width of 0,1 to 3,0 m/s. A cylindrical filtrate circuit at the other end of the cell drains off the displaced contents of the pores. The displaced contents of the pores and the filtrate are collected in a cup-glass (16). This collecting container is on a digital balance (17), so that the cumulative quantity of filtrate can be determined depending on the duration of the experiment.

The second component of the experimental machine is the circulation unit (mud container and the piston pump) forming a closed circulation. The Duplex-piston pump circulates the fluid from the reservoir (1) past the front side of the rock specimen and back to the store container. The pressure shocks in the circulation are reduced by a pulsation damper (9). In order to achieve a certain system pressure, the autoclave container can be impressed with a pressure up to 10 MPa over a gas bottle (nitrogen) (19) with a pressure-keeping valve. After the experiment, the pressure is reduced over an outlet valve (20). In the pressure container, two heating systems allow the heating of the fluid to the requested temperature.

In the oil circulation feeding the heat exchanger in the Hassel's cell, a heating coil (3) is integrated in the interior pressure container; further, there are electric heating wires (2) in the walls of the mud autoclave. The adjustment of the temperature is done over a thermo control with sensor (4).

The control of the temperature is made possible by a digital display (18). After the experiment, the drill mud and/or the treatment fluid can either be circulated out or drained off over a valve (6) at the pressure container.

Through a new development (Encl. 24 B) it is for the first time possible to apply counterpressure to the pore contents according to the well conditions at this machine.

In order to realise the counterpressure, the fluid store container (1) is at first linked to the counterpressure side over a ventilation valve

(3) to fill the Hassel's cell with fluid. Then, the requested counterpressure is applied through pressure application of a second fluid store container (9) with nitrogen. The nitrogen also ensures stabile pressure conditions during the performance of the experiment through a pressure keeping valve (6).

Viscosity Measuring Instruments

In order to determine the rheologic properties of the drill muds and treatment fluids to be used, a Fann- viscometer, model 35 S, can be used. This is a rotation viscometer which works according to the Couette-Hatschek-principle (Encl. 25 A). The outer cylinder (1) can rotate over a gear (4) with six different rotation frequencies. This makes the registration of the flow curve in a shear gradient area up to 960 s-1 possible. The induced shearing stress on Rotor is transferred to the interior, coaxially positioned body of revolution fixed with a torsion spring (T7), which is travelled out of its position of rest, over the drill mud and/or treatment fluid being in the ring gap. The shearing stress of body of revolution effective at the interior body of revolution (2) is a function of the tenacity of the fluids and the shearing stress on Rotor. The twist ling angle of the interior body of revolution is read from a linearly graduated scale in scale gradations /223/.

In order to be able to test fluids for their rheologic properties at temperatures up to 94°C, a mud container with electric sheath heating is at disposition.

In order to determine the viscosity of crude oil, the Haake-rotation viscometer RV 100 (Encl. 25 B) can be used.

Porosimeter

To determine the pore radius division of the model rocks, the Erba-pressure porosimeter, model AG/65, is used (Encl. 26). The instrument built in a metal cabinet essentially consists of /224/:

a mercury level measure and registration system,

a pressure intensifier as well as

a pressure measuring and registration system

Fluid Permeameter

To determine the permeability of the core specimen, a fluid permeameter is used (Encl. 27). The core (3) is built in a Hassel's cell and enclosed by a rubber cuff (2). Through sheath pressure (600 - 2500 kPa), the cuff is pressed close to the specimen, so that random currents can be excluded. For Bentheim sandstones, a maximum differential pressure of 5-10 kPa is reached alongside the core. With low permeable cores the fluid pillar must be applied by compressed air (10) to adjust a differential pressure of 500-800 kPa. The volume current is determined with the help of a measuring cylinder (12); the time passed since the begin of the filtration is registered with a stop-watch.

Gas Permeameter

For registration of the effective gas permeability of caliumchloride-saturated Oberkirchen sandstones, a gas permeameter is used (Encl. 28).

Scanning Electron Microscope and Polarisation Microscope

The principle of scanning electron microscopy (Encl. 29) is based on line scanning of the surface of the specimen in the specimen chamber (2) through a primary electron beam with a deflector (1). The back-scatter and secondary electrons are recorded and lead to a visualisation of the specimen surface, which is projected to a screen with the help of a video intensifier. The scanning of the specimen surface takes place under vacuum, so that the electron beam will not be scattered by air molecules /220, 221, 222/. Polarisation Microscope encl. 30.

21. ANALYSED DRILL MUDS AND TREATMENT FLUIDS

Drill Muds

Out of the multitude of different drill muds, the following shall be analysed within the scope of this thesis used for the book:

- dehydrill HT-muds,
- bentonite muds
- chalk muds as well as
- inhibited and/or non-inhibited muds

These drill muds can be divided into those:

- being poor in solids
- containing solids

Concerning their composition, these drill muds have to be conditioned for coping with the following tasks:

- Quick formation of a sealing filter cake

Mastering this task requires a short mud-spurt-phase and low filtrate losses, and the following factors have to be taken into consideration:

- grain size distribution of the solids within the mud,
- pore radius distribution of the formation rock and
- share of the bridge-forming solids of the all over contents of solids in the muds

- Solubility of the filter cake in acids before start of production
- Minimization of the formation damage by the filtrate

Since suitable bridge-forming solids (sort, grain size, concentration, solubility) quickly seal the formation in the area near the well and thus are able to reduce the formation damage to a minimum, clay and chalk are used as solids in these muds. In order to quickly seal the pore spaces of the Bentheim and/or Oberkirchen- sandstones, chalks of different grain sizes (see encl. 62-65) with a maximum share of 17,8 % of the weight (see encl. 20) is used.

Treatment Fluids

The frac- and gravel formation fluids are determined in co-ordination with experts of the industry.

Versagel (water-basic fluid),

MY-T-Oil (oilbasic frac-fluid),

Hydropac (water-basic gravel formation fluid) and

HCl and/or FHF/HCl

shall be tested.

22. USED MODEL ROCKS

The most important bed-technical parameters of the rock in view of formation damage are permeability, porosity, and pore radius distribution /225 - 229.

Influencing factors on the rock properties result from the kind of deposition of the storing rock, compaction and formation of minerals. The meaning of clay minerals in the pores of the rock shall be emphasised here, since they can influence the formation damage decisively /134, 218, 219/. The petrophysical parameters of the used model rocks can be seen from table 4.

The varying permeability values of the used sandstones can be attributed to the fact that the corresponding cores would have to be extracted from different profiles (vertical, horizontal) /217/.

Table 4: Petrophysical of the used rocks

Gestein	Quarz in %	Feldspat in %	Feinpartikel in Poren in %	Opak-Minera-lien in %	Schwer-minera-lien in %	Ton in %	Poro-sität in %	Permea-bilität in µm² (D)	Poren-größe in µm
Obern-kirchner Sand-stein	92	3,5	1,0	0,75	0,75	2,0	15-20	0,005-0,02 0,015-0,03*	15±10
Bent-heimer Sand-stein	94	1,5	1,5	0,5	0,5	1,0	20-25	1 - 3,5	25±10

Bentheim Sandstone

The used Bentheim sandstone is assigned to the middle-valendium period (Cretaceous). The structure of Bentheim sandstone is characterised by well-sorted sandstone layers. The sorting coefficient is at 1,2 to 1,4.

The media value of the grain diameters is at 0,2 mm. The grain curve is only moderate, which indicates relatively short transportation distances.

The porosities of the productive horizons consisting of Bentheim sandstone are at 20 - 25 % on average, the permeabilities at 1 - 3,5 µm² (D). Iron occurs in the form of siderite (iron carbonate); during weathering, the iron dissolves and falls out again into the rock as oxide and hydroxide in form of crusts and skins. Through fall-out of hydroxide, locally solidified areas arise, since the iron components take over the function of a binding agent. If the iron contents increase, concentric-dished structured crusts of ochre-yellow colour occur /42, 218, 219/.

The Bentheim sandstone mainly consists of quartz (at 94 %, coarse-grained) and does hardly contain any rock fragments. In the pore spaces, up to 1 % of the weight of clay minerals (kaolinite, illite) can be found /42, 62, 218, 219/.

Oberkirchen- Sandstone

The specimen origin from the Oberkirchen quarry in the southern part of the Bückeberg. They are compact and fine-grained deposits of the Cretaceous (Wealden) /116 - 118/. The average porosity is 15 - 20 %. The fluid permeability (5-20) * 10 -3 µm² (5 - 20 mD). Concerning its mineralogical composition, the Oberkirchen- sandstone is different from Bentheim sandstone owing to its small share of mica, epidot, circone, tourmaline and ores. The pore spaces are partly filled with clay minerals (<2 %). These rocks are extensively compact and almost homogeneous.

23. DISCUSSION OF THE RESULTS

Carrying out of the Experiment

To get reproducible results, certain mixing rules were kept when preparing the drill mud and the treatment fluids.

With the DHT-mud, a 2,5 % of the weight of master solution was made with tap water. The initial mixing was made with the help of a blade mixer with a rotation frequency of 800 min -1. The stirring lasted 24 hours. Then, a 72-hour-swelling took place in a state of rest.

Before being diluted, the master solution was stirred for 5 minutes with the blade mixer. For dilution, tap water was added to the master solution in a necessary amount; the mixture was then homogenised with the blade mixer at 800 min -1 for 2 hours; then, a period of rest of 24 hours followed.

Before each use, the specimen was stirred with a blade mixer at 800 min -1 for ten minutes.

Bentonite muds were prepared 8 % of the weight of and stirred for 24 hours; then, a period of rest of 3 days followed to achieve an extensive swelling of the clay.

The polymer solutions were stirred for 24 hours in order to ensure an extensive hydration of the polymer.

Solids were stirred with the above-mentioned suspensions for six hours to ensure a homogenous distribution of these substances in the suspension.

Under treatment fluids, the preparing rules for VERSAGEL, MY-T-OIL, HYDROPAC and ACIDS shall be indicated here.

To prepare VERSAGEL, a saturated salt solution of pH 7 was mixed with 4,8 g/l WG-11 (HPG), so that a pH value of approximately 8,8 appeared. A following reduction of the pH value to 5,5 - 6 through addition of 40 %-HAC or respective hydrochloric acid leads to a complete dissolution of the polymers in the liquid phase.

After this treatment, the pH value was once again increased to 6.5 - 7 through addition of K-34 (Natrium carbonate).

To allow the formation of a network, 0,4 cm^3/l CL-11 (titan-cross linkers) were added. The so-called "lepping-time", a unit for the cross-linking-time, is at about 2 minutes here. Directly before the use, a breaker (natrium- or ammoniumperoxydsulfate) of a concentration of 0,006 to 0,06 % of the weight for use at 93°C and of 0,024 to 0,06 % of the weight for use at 60 °C was added to the gel.

To prepare MY-T-OIL, 250 cm^3 diesel oil and 2 cm^3 MO-55A (gel agent) were mixed with each other; then, K 34 was added. After that, a slow addition of 0,625 cm^3 MO-56 (Natrium aluminate solution) as catalyst took place before the use. With slow addition of this catalyst, a gel with a pH value of 7 develops. In case this value is considerably fallen below or exceeded, the gel is destroyed.

HYDROPAC was prepared by adjusting saturated salt water to a pH value of 3,5 through addition of 40 % HAC or corresponding hydrochloric acid. Then, 9,6 g/l WG-17 (HEC) were added. If the polymer was completely dissolved after 10 minutes, the pH value was increased to 8 - 10 with the help of NaOH. A fifteen-minute-stirring followed; finally, the pH value was once again reduced to 6 - 7.

For acid treatment, hydrochloric acid C(HCl) = 4 mol/l or mixtures of hydrochloric acid C(HCl) = 3,2 mol/l and hydrofluoric acid C(HF) = 1,5 mol/l were used.

Filtration Experiments

Within the scope of the present Author thesis, static as well as dynamic filtration analyses were carried out.

Examination of Static Filtration

With static filtration, the mud is in rest. The static filtration behaviour of drill muds and treatment fluids used was examined with the help of the well simulation circulation machine.

In order to get meaningful test results, certain drill muds and treatment fluids were assigned to core specimen of permeability as similar as possible. Before each experiment, the pressure container of the circulation machine was filled with 2,4 l of the respective fluid; the experiments were carried out under corresponding temperatures and pressures. The sheath pressure applied to the rubber cuff was always about 2,0 MPa above the respective differential pressure. The heating of the rock specimen was affected through an oil heating over the sheath of the Hassel's cell; the heating of the fluid in the pressure container through an electric sheath heating as well as through an interior heating coil flown through by oil.

The experiment lasted 1 - 7 hours. The selected drill cores were built in the Hassel's cell directedly; thus, it was ensured that the flowing through of the core always took place in the same direction in the circulation machine and/or the permeameter.

After building the core in the Hassel's cell, fluid from the fluid store container was pressed into the circulation with the help of nitrogen; through this single process, the fluid could flow back into the fluid store container past the front side of the rock specimen.

So, the Duplex pump was not put into operation with static experiments.

Examination of Dynamic Filtration

A dynamic filtration is present if the fluid circulates. The dynamic filtration behaviour of the used fluids was examined in the circulation machine as well with the same experimental set-up. The requested velocities ($\leq$ 1,5 m/s) are adjusted at the flow control. The maximum circulation temperature was 90 °C during the filtration. The maximum differential pressure was adjusted to 6 mega pascal (MPa). A duplex pump circulated the fluids past the front side of the core. In order to record the influence of repeated composition and decomposition of the filter cake (alternating static and dynamic filtration) on the formation damage, it was filtrated for 4 hours, 1 hour dynamically and one hour statically by turns, in a corresponding experiment.

Examination of Model Rocks

The sandstone cores of a diameter of 2,45 cm were sewed into 25 cm long pieces after being drilled from the rock. After cleaning them from sticking soilings, the cores are dried in a drying cabinet at about 110 °C. Through weighing the core after different durations of staying in the drying cabinet, it could be proved that at the latest after 48 hours a mass constant sets in.

After inscription, the cores were soaked in an exsiccator, where the specimen was at first evacuated for at least 12 hours to achieve a saturation as complete as possible of a 1,5 % of the weight of KCl-solution. The KCl solution should prevent a swelling of the clayish particles of the sandstone, so that its original permeability was maintained.

The clay-inhibiting effect of the KCl solution is caused by the fact that, according to MONDSHINE /5/, calium ions have a relatively small diameter and, according to BLACK & HOWER /5/, substitute other exchangeable cations of the clay; therefore, they can easily enter between the single clay particles and restrict the swelling owing to a weak hydration tendency of the occurring K-clays (encl. 31).

Permeability measuring's were also carried out with cores having successively been soaked in distilled water, tap water and 0,5, 1, 1,5, 2 and 2,5 % of the weight of KCl solution. These measuring's were also carried through with dissolution of the KCl solution in the pore space from 2,5 to 0 % of the weight. During that, the specimen stayed in the next higher / lower concentration degree for 10 hours each time. Then, a penetrating with 20-time pore volume followed. The permeability was measured after each saturation degree.

Porosity and Pore Radius Distribution

The effective porosity was determined with the help of the soaking method /119/. The dried core pieces were weighed and soaked with methanol in an exicator. After removal of the sticking soak fluid, the mass of the soaked specimen was determined.

The effective pore volume is calculated from the mass difference between the soaked and the dry core related to the density of the penetrated fluid. Then, the soaked core pieces were weighed under methanol.

If this value is subtracted from the mass of the soaked specimen in the air, the mass of the displaced fluid is obtained. The all over volume of the specimen could be calculated accordingly with the density of methanol. The quotient of effective pore volume and all over volume results in the effective porosity of the rock.

The determination of the pore radius distribution was made with the Carlo-Erba-mercury-pressure-porosimeter.

Each time, a core piece of app. 4 - 5 g was brought into the glass dilatometer. After filling the dilatometer with mercury and evacuation of the specimen, the core piece was gradually applied by ventilation in degrees of 20 kPa up to the atmospheric pressure; the corresponding mercury grade was read from the dilatometer neck. The filled dilatometer container was then inserted into the autoclave of the porosimeter. The volume of the mercury having penetrated the pores of the core under pressure load was recorded as function of the adjusted pressure (max. 100 MPa).

Pore radius of 75 nm to 7500 nm can be registered with the low-pressure part, those between 7,5 nm and 75 nm with the high-pressure part of the device.

The following equation shows the relation between adjusted pressure p and pore radius r:

$$r = \frac{7500}{P} \qquad \text{Eq. 48}$$

r = pore radius in nm.

P = adjusted pressure in MPa.

Permeability

The permeability k can be determined according to the law of DARCY:

$$K = \frac{0.1 * \mu * Q * L}{A * \Delta P} \qquad \text{Eq. 49}$$

K = Darcies

μ = viscosity of the penetrating medium [CP]

Q = penetrating rate [cm³/s]

L = length of the core [cm]

ΔP = differential pressure [MPa]

A = cross section of the core [cm²]

If the pores of the rock contain two or more media, the DARCY equation with stationary current results in an own permeability for each fluid, the so-called "effective permeability" ke.

The relation between effective and absolute permeability is shown by the "relative permeability".

$$kr = \frac{ke}{k} \qquad \text{Eq. 50}$$

The DARCY equation in its present form cannot be applied to compressible media. Enclosure 32 explains the viscose current valid for fluids and the glide current prevalent in the λ-zone. The high velocity of the gas molecules in the λ-zone contradicting the theory of viscose current explains the differing behaviour of compressible media.

Through graphic tracing of the permeability values ka measured under different pressures on l/pm, the equivalent fluid permeability of the core for the corresponding gas can be determined through extrapolation of the marked straight line with (l/pm) = 0.

The original permeability ku of the rock specimen was determined with the fluid permeameter described in 20 under EXPERIMENTAL EQUIPMENT. Therefore, the soaked cores were built in the Hassel's cell. The rubber cuff was applied by pressures up to 2500 kPa. In order to ensure a sufficient flow rate, a differential pressure of 10 kPa for Bentheim sandstone and of 500 kPa for Oberkirchen- sandstone was applied. After adjusting stationary flow conditions, the volume of the fluid flowing through the specimen within 180 s (Bentheim sandstone) and 5 - 20 min (Oberkirchen- sandstone) was measured. Therefore, a balance and a measuring cylinder with an accuracy of 0,1 cm^3 were used. In order to increase the reliability of the results, at least two measuring's were carried out each time and the average calculated.

The effective permeability for oil was also measured on the core specimen of Bentheim sandstone. Therefore, the KCl-solution in the core was displaced with crude oil from the bore-hole Eddesse North until water saturation was achieved.

Besides the fluid permeability (KCl-solution), the effective gas permeability Ke was determined with the gas permeameter described in 20 for Oberkirchen- sandstone.

The measuring of permeability of damaged cores took place analogously. While the measuring of cores was affected directly after damage through drilling, especially frac-fluids needed a waiting time of 24 hours in the autoclave, so that the effect of the gel-breaker could completely enforce.

Only after that, the measuring's were carried through in the Hassel's cell. Through the formation damage caused during filtration, a higher differential pressure has to be applied to obtain measurable values.

REM-experiments

A special preparation of the specimen was necessary for the REM experiments. Therefore, 3 - 5 mm thick rock lamellas were sewed out of the front side, the middle and the end of the core and cut up.

The specimens were deeply frozen in two phases with the help of the shock freeze technology. In the first phase, the specimen was shock frozen to -160 °C in a metal container filled with deep cold 2-methylbutan.

In phase 2, the specimens were cooled down to -196 °C in liquid nitrogen.

Now, the specimen was once again broken in the coolant. The specimen pre-treated this way were then freeze-dried.

Freeze drying means the withdrawal of microcrystalline solidified water from frozen material.

The drying was affected avoiding the liquid state of aggregation through sublimation. The sublimation takes place in under vacuum; the temperatures of the specimen brought in have to be below -10 C.

The specimen had to stay at least 24 hours in the facility, according to size and form as well as mass of free and bonded water and the relation between fresh and dry substance.

The freeze drying comprised:

Initial drying phase with sublimation on the surface of the specimen at appr. -75 °C main drying phase at a temperature of appr. -20 °C.

After freeze drying, the specimen - still under vacuum - was slowly warmed up to the surrounding temperature.

The increase of the pressure in the freeze drier to the surrounding pressure had to be affected over several hours to exclude changes in the structure of the specimen because of sudden pressure load.

After freeze drying, the specimen was glued to Al-plates with the special adhesive Leit-C and dried for 24 hours in an exicator. Then, gold was evaporated onto the specimen; after that, the examination with the scanning electron microscope was carried through.

Thin Grinding Analysis

Generally, thin grindings have to be prepared very carefully, so that the pore space relations are not distorted and no serious mistakes will occur during the analysis.

For preparation, small grind blocks are sewed out of the concerned specimen and treated at the surface.

After a sufficiently even surface has been obtained, the specimen piece is glued to the object slide with special means (Arochlor, Canada balm).

After that, it is sewed once again and the second side is grinded; finally, the cover glass is fixed with the same adhesive. The thickness of the thin grinding preparations is at 25 µm.

In order to avoid breaking during the work process, especially porous rocks have to be impregnated with artificial resin 114/. This is also done especially with examinations of the pore spaces, where a blue resin contrasting the rock is used.

To examine the damaged cores, the 25 cm long specimen were divided into five, every section divided in half, and then dried at 60 °C.

The following impregnation with the special blue resin was done under vacuum in order to avoid formation of bubbles. Then, the thin grindings were produced, which could be tested for contamination with solids and/or polymers with the polarisation microscope (encl. 30).

The filter cake as well can be identified and evaluated in thin grinding.

Chemical analysis

The filtrate penetrating through the rock cores can be analysed for ion concentration and possibly for certain tracer ions with the help of atomic absorption spectrometry AAS (encl. 33). Within the scope of the tests carried out, Na+ served as tracer to determine the amount of mud filtrate, and K+ as trace element to determine the driven-out volume of bed water.

Further examinations

Further, the so-called EASY Particle Sizer 3600 E by the company Malvern Instruments is used to determine the particle size of microsoil, texchalk, Calcidar 40, DS-20 and drill particles.

23 DISCUSION AND RESULTS

In order to ensure a clearly arranged structure of this paragraph, the results having been worked out shall be discussed under the following titles:

Impact of salinity and pH-value of penetrating water on the permeability of the core specimen, damage of examined core specimen through drill muds, damage of the core specimen through frac- and gravel carrier fluids, change of the level of damage degree of the core specimen through re-extracting attempts as well as change of permeability of examined core specimen through acid treatment.

Impact of salinity and pH-value of the penetrating water on the permeability of the core specimen

In order to determine the degree of damage of core specimen, permeability measuring's are carried out; gaseous or liquid media can be used for that.

When fluids are used as measuring medium, the fluid composition plays an important part to obtain reproducible results. Since the used Bentheim and Oberkirchen- sandstones contain clayish components (see table 4), the impact of salinity and pH-value of the measuring fluids on the change of permeability of the treated sandstones had to be determined. Enclosures 34 and 35 show the change of permeability of examined cores depending on the KCl-concentration of the measuring fluid.

It is obvious, that the permeability of the core specimen increases with increasing KCl concentration of the measuring fluid. With Bentheim sandstone, the permeability stays constant from a KCl concentration of 1,5 % of the weight on, while it diminishes slightly with Oberkirchen- sandstone. When solutions of decreasing KCl concentration were used for measuring the permeability of these sandstones, the permeability of the core drastically diminished at KCl concentrations lower than 1,5 % of the weight. It is further acknowledged, that the permeability values measured with solutions of decreasing KCl concentrations are considerably higher than the permeability values measured with solutions of increasing KCl concentrations, especially with Bentheim sandstone. As described already under 22, salt solutions lead to a decomposition of the hydrate sheath of the clays participating in the composition of the rocks. The decreasing thickness of the hydrate sheath leads to partly released cross sections of the pore channels, so that higher permeability values can be measured. If the salt concentration of the measuring fluid is further increased, a cation exchange can take place at the basic planes of the particles. Through treatment with a KCl solution, for example, a K-montmorillonite being considerably less productive develops from Na-montmorillonite. According to /5/, the higher permeability values with using measuring fluids of decreasing salt concentration have its cause in the fact that the solutions of higher concentration used at first lead to a cation exchange at the basic planes of the particles. Due to the fact that K-montmorillonite has a considerably lower ability of swelling than Na-montmorillonite, hydrate sheaths of the particles being much narrower are realised in the area of higher concentration, so that permeability values being considerably higher can be measured.

In enclosures 36 - 38 is documented to what extent cation exchange and concentration decline take part in the change of the measured permeability values.

From enclosure 36 can be seen, that the penetrating of a core specimen soaked in a 3 % of the weight of NaCl-solution with a NaCl-solution of the same concentration (unchanged concentration decline, no cation exchange) does not cause a change of permeability values. If a penetration with deionised water is carried out subsequently, the hydration degree of the clays increases owing to the increasing concentration decline and the permeability values increase drastically through narrowing of the cross sections of the pore channels. Should the core specimen be soaked in solutions of bivalent salts, the penetration with deionised water, in spite of an increasing concentration decline, does not cause a change of the permeability values (comp. encl.

37). If then a flooding with 3 % of the weight of NaCl-solution is carried out, a cation exchange (Ca-montmorillonite changes into Na-montmorillonite) takes place, so that a subsequent flooding with deionised water leads to a drastic permeability loss.

As is shown in enclosure 39, the spontaneity of the change in concentration of a salt solution has decisive influence on the degree of the damage of the flooded core specimen. While a gradual reduction of the salinity does not cause a permeability reduction, a spontaneous reduction of salinity from 1,5 % of the weight of KCl to 0,1 % of the weight does affect a drastically reduction of permeability (increase of the DR-value from 0 to more than 80 %). Through gradual reduction of salinity, the resulting dispersed clay particles from the pore wall in the transport fluid are minimal, so that a permeability decrease through formation of bridges at the narrow places of the pore channels can be excluded 112/. With spontaneous salinity decrease, a lot of clay particles dissolve from the pore wall synchronically, so that the narrow places of the pore channels can be bridged during their further transport; the permeability of the core specimen considerably reduces consequently.

Enclosure 40 shows the degree of the damage (DR) depending on the pH-value of a freshwater bentonite-mud. It is obvious that the damage degree increases with increasing pH-value.

Higher pH-values lead to an intensified hydration of the clays dispersed in the mud on one hand. On the other hand, the higher pH-value of the filtrate of these muds causes a higher hydration degree of the clayish components of the rock. Through a more effective bridging of the pore-holes (increase of the hydration degree of the clays in the mud) and narrowing of the cross sections of the pore channels of the rock through higher hydration of its components, the permeability of the specimen reduces with higher pH-values.

In enclosure 41, the changes in permeability of a core specimen of Bentheim sandstone soaked in a 1,5 % of the weight of KCl-solution depending on kind and amount of the penetration medium are described. It is obvious that the specimen did not change after penetrating with a 1,5 % of the weight of KCl-solution even after 20 PV. During the subsequent penetration with tap water, the permeability reduces to a minimum of 0,5 μm^2 (D) after further 20 PV.

In enclosure 42, the changes of the pH-values of the filtrate with decreasing calium content are described.

In view of the examinations of the degree of carrier damage and sectional damage ratio to be carried out within the scope of the Author thesis, it can be derived from the results of the experiments being described in this paragraph, that the permeability measuring's with liquid media have to be carried out under determined conditions to obtain reproducible values. The following boundary conditions have to be maintained:

a) The core specimen have to be saturated with 1,5 % of the weight of KCl-solution.

b) A 1,5 % of the weight of KCl-solution shall be used to measure the permeability of the core specimen.

c) Spontaneous changes of salinity in the pore space are to be excluded.

d) Changes of pH-value have to be avoided.

Damage of examined core specimen through drilling-muds

The results of the experiments in view of the damage of Bentheim sandstone through DHT- and bentonite-muds are described in enclosures 43 – 48. An account of the DR-values (comp. 2.1.2.1) was left out intentionally. The cause is that the permeability of the non-segmented, damaged core is mainly determined by the intensity of the damage in the first section of the core. So, the SDR-values having been determined for the first core-segments are identical with the DR-values, so that the description of the last mentioned key data can be left out without any loss of information.

In enclosure 43, the damages caused by DHT- and/or bentonite-muds with a circulation temperature of 90 °C, a differential pressure of 3,5 MPa and a circulation velocity of 0,6 m/s over a circulation time of 1 hour are shown. It is obvious that a 4% of the weight of bentonite suspension damages the examined core the least. The SDR values are reduced from 80 % in the first segment to appr. 25 % in the last segment. Is the core loaded with a bentonite-mud treated with Hostadrill 2825, the degree of damage increases to almost 100 % in the first segment, which hints at a better sealing of the outer filter cake produced by this mud. The improved sealing of the filter cake can be assigned to the fact that the long-chain Hostadrill-molecules close the gaps between the particles in the cake-connection; thus, the permeability of the core segment is further reduced. It is, however, disadvantageous that the filamentary polymer-molecules having been swept into the rock during the "mud spurt phase" are carried further into the rock with progressing filtration, so that, through blocking of the narrow places of the pores through these polymers, the SDR-values of corresponding segments increase (comp. encl. 44 and 45 pictures for the same cores).

Treatment of the bentonite-mud with short-chained filamentary polydrill-molecules leads to the reduction of the SDR-value of the first segment to 60 %, while the SDR-values of the subsequent segments are identical with those after the damage through the bentonite-Hostadrill-mud.

The damage of the core through DHT-muds, no matter whether pure DHT-muds or muds treated with polymers, (comp. encl. 43) leads to a complete damage of the whole core (SDR = 90 %). The reason is that the DHT-lamellas are, unlike bentonite, easy to deform /147/. Under corresponding differential pressures, the DHT-lamellas are pressed through the relatively wide pore holes of the Bentheim sandstone and transported through the pore channels (comp. encl. 46 – 48). A sealing outer filter-cake is missing. The transport of the DHT-lamellas as well as of the filamentary polymer-molecules effects a bridging of the narrow places of the pore channels in the core segments, so that their SDR-values increase.

In Figure 44 encl. 39, the influence of an electrolyte load of DHT and/or bentonite-muds on the level of damage of the examined Bentheim sandstone is described. Through loading these drilling-muds with CaCl2, the clay particles as well as the polymers are to a considerable extent dehydrated. Through separation of the bondings between the clay particles and the filamentary polymer molecules and coagulation of the clay particles, larger agglomerates are formed, which better bridge the pore holes of the Bentheim sandstone to be examined. Consequently, the SDR-value in the first segment rises. The partly dehydration of the filamentary polymer molecules leads to the fact that the blockage of the pore channels of the rock stays incomplete in the single segments, so that their SDR-values strongly diminish (comp. encl. 49 and 50). From encl. 44 can be seen that the SDR-value of a bentonite-polydrill-mud loaded with CaCl2 reduces from 80 % in the first segment to appr. 20 % in the last segment.

The SDR-values of the core specimen of Bentheim sandstone treated with drilling-muds having been exposed to a temperature load of 90 °C for 120 hours are described in enclosure 40. It is obvious that the SDR-values with damage of the core through DHT-muds are higher than those with bentonite-muds. The causes of the damage of the examined sandstone cores here are the same as were described in context with the results shown in enclosure 39. In the present case, the effects are, due to the lasting temperature load of the mud, more serious.

In encl. 46, the SDR-values after loading the core specimen with DHT- and bentonite-muds, which contain PAC as well as Polydrill and/or Hostadrill as polymers, are outlined. Here too, it is obvious that the SDR-values decrease rapidly with damage of the core through bentonite-muds, depending on the penetration depth of the damage, while these are above 90 % for loading he core with DHT-muds from the first to the last segment.

The improvement of the mud properties in view of the formation protection through addition of Microsoil can be made clear with the help of the results of the experiments shown in encl. 47. While with loading the core with bentonite-mud, the SDR-value decreases to less than 10 % in the last segment through addition of Microsoil, a considerable reduction of the SDR-values takes place with DHT-muds as well. This can be assigned to the fact that the DHT-mud in combination with Microsoil also causes a more effective bridging of the pore holes of the rock, so that the further transport of solids in the pore channels of the rock is restricted (comp. encl. 51 and 52). The polymer- and DHT lamellas penetrated into the pore channels during the "mud spurt phase", however, are transported through the filtrate, so that they settle at the narrow places of the pore channels and cause a permeability barrier there.

The excessive SDR-value in the last core segment hints at an accumulation of DHT-lamellas as well as of polymers. This assumption is confirmed by a REM-exposure (comp. encl. 46). In enclosure 43, the SDR-values of the core loaded with bentonite- or DHT-muds with or without temperature pre-loads and/or with or without addition of electrolyte are shown.

It is obvious that the core damage is less with a temperature pre-load of the used mud than with a CaCl2-load of the mud. While, for instance, the SDR-values reduce from initially 55 % to 10 % in the last segment, they are above 30 % over the whole length of the core with damage through CaCl2-contaminated muds.

According to the details of the company, the formulae stated in encl. 49 are formation -protective muds. These muds were tested in view of formation damage. The results are shown in encl. 50 – 57. It is obvious that, among the muds no. 1 to 15, muds no. 10 and 11 damage the formation the least (comp. encl. 52). They are clay-free muds containing XC and HEC as polymers and texchalk as solids. Mud no. 10 also contains gypsum, while caliumacetate was added to mud no. 11. The worst result was obtained with fluid no. 1, a clay-polymer-mud treated with K-acetate. The SDR-value here decreases from initially 75 % to only appr. 55 % in the last segment. Obviously, the low clay concentration of this mud is not enough to quickly bridge the pore holes of Bentheim sandstone, so that the solids and polymers washed into the rock during the "mud spurt phase" lead to a reduction of the permeability of the core in all segments during the transport through the following filtrate.

The far best result was obtained with fluid no. 16 (encl. 56). Here, the SDR-value of initially appr. 50 % reduces to appr. 2 % in the first segment already. In the last segment, the core is not damaged. With view to this good result, further experiments were made with this clay-free mud (comp. encl. 56 and 57). It is obvious that the damage is minimal with cores saturated with KCl-solution as well as with oil-saturated cores. It can further be stated that the change of the saturation relations in the pore space leads to an increase of the SDR-values in the examined rock (comp. The curve shown in encl. 56 for the oil-saturated core to that of the KCl-saturated core). It can further be seen from encl. 56, that a load of 120 hours under a temperature of 90 °C leads to a slight increase of the SDR-values alongside the core.

Fluids no. 1, 5, 6, 10 and 14 were tested concerning the damage of Bentheim sandstone after a load of 120 hours under a temperature of 90 °C; the results are shown in enclosure 49. According to this description, the best results were obtained with fluid no. 10, a clay-free drilling-mud containing texchalk as solid and gypsum as salt. It can be seen that the SDR-value reduces to 0 already in the fourth segment.

In encl. 57, the SDR-values are shown depending on differential pressure. With initial pressures of 2 or 3 MPa, the counter pressures were adjusted in a way that each time, a differential pressure of 2 or 1 MPa was realised. It can be seen from encl.57, that the SDR-values alongside the core are at 70 % with a differential pressure of 1 MPa. The SDR-value in the first core segment is higher with a differential pressure of 2 MPa than the

SDR-value with a circulation pressure of 1 MPa. This indicates that higher circulation pressures lead to a compaction of the filter cake, so that the permeability of the first core segment decreases correspondingly. This tendency is confirmed by encl. 58 and 59. It can be seen from these enclosures that the SDR-value tends towards 0 already in the fourth or fifth segment with a circulation pressure of 2 MPa.

It can only be assumed here that, conditional on the intensified compaction of the filter cake, the washing of solids into the rock is restricted to a minimum, so that the last segments are not influenced by that.

It can be seen from the last three enclosures, that there is also a damage of the specimen under a differential pressure of 0 in the first segments. The reason is that, as stated in 7, the osmotic and capillary pressure are effective as driving force of filtration besides the differential pressure. If the osmotic pressure is eliminated, as to be seen from encl. 59, the SDR-values are considerably lower. The penetration depth of the damage is limited to the first segment in this case.

In encl. 60, the dependence of the degree of damage with core specimen of Bentheim sandstone on the duration of circulation of the used muds is shown. It can be seen from the enclosure that the DR-value increases with the duration of circulation and reaches appr. 100 % after 7 hours.

The increasing DR-values are mainly to be attributed to the fact that the hydration degree of clayish components of the rock increases with displacement of monovalent K-ions of the pore space by the filtrate of freshwater-basic muds, so that the effective cross sections of the pores are reduced.

In encl. 61, the damage degree of Bentheim core specimen depending on the clay concentration in drilling muds these cores have been damaged by is shown. It is obvious that the DR-values decrease with a DHT-concentration of up to 2 % of the weight and with a bentonite concentration of up to 3 % of the weight. While the increase of the bentonite concentration in the drilling mud leads to a quicker bridging of the pore holes of the rock and the formation of a more compact filter cake, the decrease of the damage with increasing DHT-concentration takes place through the intensity of its gelling-tendency.

The results of the experiments on the damage degree with Oberkirchen- sandstone through loading with different drilling-muds are described in encl. 62 – 67. It can be seen from encl. 62 enclosure 57, that the SDR-values decrease from initially more than 70 % to less than 30 % in the last segment. This tendency, which could not be stated with Bentheim sandstone, has to be attributed to the fact that the deformable DHT-lamellas settle at the narrow places of the pore channels of Oberkirchen- sandstone already being narrow, and thus lead to a sealing of the corresponding segments. The restriction of the further transport of the DHT-lamellas dependent on that causes even lower SDR-values in the subsequent segments. It can further be seen from this enclosure, that higher differential pressures lead to an intensified damage under the same conditions apart from that.

The results described in encl. 63 emphasise the statements made in view of encl. 62. Additionally, it can be seen from enclosure 58, that a higher temperature load of the used DHT-mud leads to the examined core specimen being badly damaged in the last segment under higher differential pressures (2,2 and/or 3,5 MPa). This hints at the fact that the possibility of further transporting the DHT-lamellas also through narrow capillaries under higher differential pressures arises through dehydration as well as delamination /147/ of the DHT-particles through a higher temperature load, so that also the last segments of the examined specimen are highly damaged through blockage of the narrow places of the pores. The statements made here can be verified with REM-exposures and thin grinding analysis. Encl. 53 – 57 REM-exposures of segments no. 1, 3, and 5. It is obvious that

DHT-lamellas in the pore space of the rock are still detectable in the last segment of the core specimen.

In encl. 64, the SDR-values of core specimen of Oberkirchen-sandstone after varying, dynamic and static filtration are shown. By turns, it was filtrated each time two hours dynamically and two hours statically. It is obvious that the SDR-values are lower with DHT-muds to which Hostadrill was added than with pure DHT-muds.

In encl. 65, the SDR-values for core specimen of Oberkirchen-sandstone being damaged with DHT-muds of different composition and a 120-hour thermic pre-load of 90 °C are shown. It is obvious that the damage of the core is strongly restricted through the protective effect of Hostadrill 2825.

In encl. 66, SDR-values having been measured after damage by DHT-muds of different composition and different electrolyte-loads are compared to each other. In this enclosure, it is especially striking that the SDR-values strongly reduce with damage through a DHT-mud having been loaded with 6 % of the weight-$CaCl_2$, depending on the segment position. This is caused by the fact that the hydrate-sheaths of the DHT-lamellas are to a great extent decomposed through bi-valent cations, so that an area-to-area-addition (A-A) of the clay particles is the consequence (coagulation).The coagulated clay-packages are no longer as deformable as it was the case with completely hydrated DHT-lamellas. Therefore, a further transport into the pore channels of the rock is no longer possible, so that the SDR-values strongly reduce in the corresponding core segments. Concerning the influence of the differential pressure and of the level of the circulation pressure on the damage of examined core specimen of Oberkirchen- sandstone (encl. 67), it can be referred to the explanations having been made with the corresponding experiments on Bentheim sandstone (see under section 23, encl. 57).

Damage of Core Specimen through Frac- and Gravel-formation Fluids

The examination of the level of damage through Gravel-formation- and Frac-fluids, whose formulae were given according to company details, was mainly affected with core specimen of Oberkirchen-sandstone. In view of the damage of Bentheim sandstone, the results of the experiments obtained with "Hydropac" are exemplarily shown in encl. 68. From this enclosure, the influence of temperature load of the used Gravel-formation-fluid and of the kind of the added breaker on the level of the caused damage can be seen. It is obvious, that, with Gravel-formation-fluid without addition of a breaker, the SDR-values decrease depending on the position of the segment and almost tend towards 0 in the last segment. This curve hints at the fact that the pore holes of the rock in the first segment were partly bridged by the components of this Gravel-formation-fluid, so that this segment shows the highest damage degree (comp. encl. 58).

Through partly bridging the pore holes of the first segment, the transport of the fluid components is impeded, so that the damage degree rapidly decreases with increasing distance to the front side of the core. If the formation fluid was treated with a breaker, a dehydration, especially of the polymers of the formation fluid, takes place depending on the time during which the breaker was effective, which goes hand in hand with a serious loss of volume /26/. The polymers diminished by the effect of the breaker can hardly bridge the pore holes of the rock, so that the damage degree decreases in the first segment (comp. encl. 59). On one hand, the influence of the breaker leads to the fact that the diminished polymers can be transported further into the core, so that, unlike a damage caused by a formation fluid without breaker, the SDR-values can increase in the subsequent segments in this case (comp. encl. 60). a temperature load of the formation fluid leads to an additional destruction of its components. The result is corresponding in view of the SDR-values. In encl. 68 – 72, the results obtained with Oberkirchen- sandstone are shown. In encl.

69 and 70, the SDR-values with treating the core specimen with the Frac-fluid "Versagel" are shown.

In view of the experimental conditions, the results shown in enclosure 64 and 65 differ concerning the level of differential pressure. From encl. 69, it can be seen that the SDR-values decrease alongside core specimen through addition of breakers to the examined Frac-fluid. The causes for the decrease of the SDR-values have already been described in the context of encl. 68.

A comparison between encl. 69 and 70 hints at the fact that higher differential pressures lead to a slight increase of the SDR-values alongside the core.

In encl. 71 and 72, the SDR-values with damage of the core specimen of Oberkirchen- sandstone through "MY-T-Oil" as Frac-fluid are shown. With this Frac-fluid, the task of the breaker is taken over through adjusting certain pH-values. From enclosure 66, it can be seen that the pH-value of the measuring medium has decisive influence on the level of the SDR-values. It has to be stated, that the degree of damage through "MY-T-Oil" is relatively low. The SDR-values decrease from originally 40 % in the first segment to almost 0 in the last segment. Higher differential pressures (comp. encl. 72) lead to an increase of the SDR-values in the first segments. The SDR-value in the last segment, in this case as well, is 0.

Change of the Level of Damage Degree of the Core Specimen with Re-extraction attempts.

A selection of damaged core specimen of Bentheim sandstone was further examined through re-extraction attempts. These experiments should clear, to what extent the solids washed into the rock could be retransported, so that the damage degree of the core specimen decreases. The results of these experiments are to be found in encl. 73 – 78.

In the following, the results shown in encl. 73 and 74 are discussed exemplarily. In encl. 73, the degree of damage is described as function of displacement time and differential pressure. It is obvious, that the damage degree, depending on the displacement time, generally decreases more or less strongly at first and that it is constant after running through a minimum or increases again and approaches a limit value. The decrease in damage degree has its cause in the fact that with reversed flowing direction, the solids and/or polymers stuck to the narrow places of the pore channels are mobilised and displaced back. Through increasing the effective cross section of the pore channels, the permeability of the core specimen increases. In case the solved solids and/or polymers stay in bigger pore spaces, the achieved permeability of the core specimen is maintained. If the solids and/or polymers are further transported, so that they can stick to different narrow places of the pore channels, the permeability decreases again; the damage degree increases. In view of the increasing permeability of the examined cores, it can be seen from encl. 73 – 78, that an increase of permeability of the core (a decrease of damage degree) can be obtained under a differential pressure of up to 100 kPa, while under higher differential pressures, a repeated increase of damage degree is to be stated. In encl. 78, the results of re-extraction attempt of oil-saturated Bentheim sandstone are shown. It is obvious, that the damage degree depending on the displacement time rarely changes here.

A possible reason why the damage degree changes only slightly through re-extraction attempts with oil-saturated cores is to be seen in the fact that, with sprinkling the solids and /or polymers with the saturation medium oil (low coefficient of friction), mobilisation and transport are possible even under low differential pressures, so that these solids and/or polymers can stick to different narrow places of the pore channels, so that the specimen is impermeable after the re-extraction attempts as ever.

Change of permeability of examined core specimen through acid treatment

Through acidification of a selection of damaged core specimen of Bentheim and/or Oberkirchen- sandstone, it should be found out to what extent their degree of damage could be reduced. The results of the acidification attempts are shown in encl. 79 – 83.

In encl. 79, the change of permeability of examined sandstones is shown. It can be seen, that the permeability of the specimen of Oberkirchen- sandstone decreases to almost the half of the original value after acidification with HCl. There are two reasons for this decrease of permeability:

1. The exchangeable cations of clayish components of the sandstone are replaced with hydrogen ions, so that a decrease of permeability of the core specimen results from change of hydration degree.

2. Through acidification of the rock matrix, fine particles are dissolved from the rock formation, so that a blockage of the narrow places of the pores is possible during their further transport.

The results of acidification experiments on Bentheim sandstone are shown in encl. 80 and 81. It can be seen from enclosure 75, that, through acidification with HCl, the SDR-values decrease alongside the core. In case a mixture of HCl and HF is used, the decrease of the SDR-values is intensified. As was described under 11 Acid, HCl mainly attacks the alkali and alkaline earth components of the matrix. Additionally, a transformation of clayish components of the rock to H-clays, which show the lowest hydration degree, takes place through HCl-treatment. Treatment of the matrix with hydrofluoric acid (HF) leads to the solution of the SiO2. That is why an additional treatment of the specimen with HF leads to an intensified decrease of damage.

It can be seen from Figure 81 enclosure 76, that the best results in view of an improvement of productivity of the rock were obtained through acidification of cores which had initially been damaged through a fluid consisting of Calcidar, PAC and K-acetate. After treatment of the core specimen with a mixture of HCl and HF, the core over its whole length can be regarded as non-damaged.

In encl. 82 and 83, the results of acidification experiments on core specimen of Bentheim sandstone are described. From these enclosures can be seen that a slight increase of SDR-values takes place with treatment of the cores with HCl. The reason can be seen in the fact that the mobilisation of finest particles through acid treatment leads to the fact that their further transport blocks the narrow pore channels of these rocks at their narrow places, so that the SDR-values increase. With treatment of the pores with a mixture of HCl and HF, SiO2 is solved as well, so that an effective enlargement of the pore channels can be assumed. Thus, the SDR-values decrease.

24. CONCLUSION

Within the scope of the present Author thesis experimental analysis on core specimen of Bentheim and Oberkirchen-sandstone were carried out in order to determine the degree of damage through drilling-muds, Gravel-formation- and Frac-fluids with circulation temperatures up to 90 °C, differential pressures of up to 6 MPa and circulation velocities of up to 1,5 m/s. Furthermore, the impact of a 120-hour temperature pre-load of these fluids on the level of damage degree should be examined. Additionally, filtration times of up to 7 hours were realised. Re-extraction attempts and treatment of damaged cores with acids should answer the question to what extent the existing permeability barriers could be surmounted.

The obtained results essentially are as follows:

1. The permeability of examined core specimen depends on the kind of soaking and measuring medium. With a 1,5 % of the weight of KCl-solution as soaking and measuring medium, the original permeability of the core is maintained best. While a gradual reduction of salinity in a core specimen soaked with a

1,5 % of the weight of KCl-solution did not cause any decrease of permeability, a spontaneous reduction of salinity from 1,5 % of the weight to 0,1 % of the weight of KCl effected a severe reduction of permeability.
The valency of the salts used in the soaking fluids has decisive influence on the change of permeability of the examined cores after treatment with other fluids.

2. Core specimen having been soaked in monovalent electrolyte-solutions (NaCl) showed an intensified decrease of permeability during their subsequent floating with de-ionised water. Were the core specimen soaked with solutions of bi-valent salts, soaking with de-ionised water did not affect any change of permeability.

3. It could be stated that the damage degree of the core increases with rising pH-value of the soaking-fluid.

4. The SDR-value in the first segment of examined cores is relatively high with loading with a bentonite-mud; in the subsequent segments, however, the SDR-values decrease rapidly with this mud. With loading the examined core specimen with DHT-muds, the SDR-value in the first segment is slightly lower than in case of a load with bentonite-muds. Here, it is of special disadvantage that the core is damaged to the same extent in the subsequent segments.

5. The treatment of drilling-muds with polymers causes a better, sealing effect of the outer filter cake; thus, the SDR-value increases in the first core segment. The transport of filamentary polymer-molecules through the filtrate, however, leads to an increased damage of the subsequent core segments.

6. A pre-load of DHT-muds with bi-valent electrolyte-solutions leads to a higher damage of the core than a 120-hour temperature pre-load under 90 °C.

7. The damage degree of examined cores increases depending on circulation temperature, circulation velocity of the drill-muds and/or the treatment fluids and on the differential pressure.

8. An addition of chalk to the used drill-muds led to a considerable decrease of damage.

9. Under a differential pressure of 0 (balanced drilling) as well, measurable damages could be proved.

10. "MY-T-Oil" caused the least damage of the used cores of the examined treatment fluids.

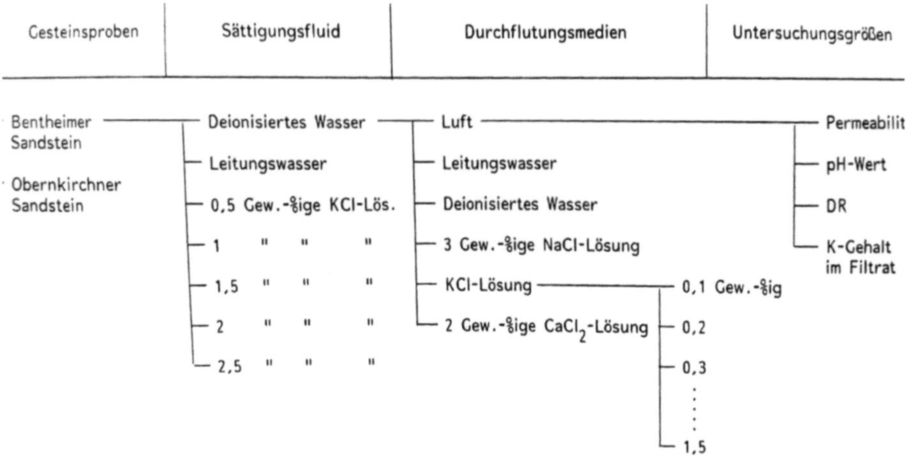

Encl. 19

The encloses from 20 to 23 not showing in the book

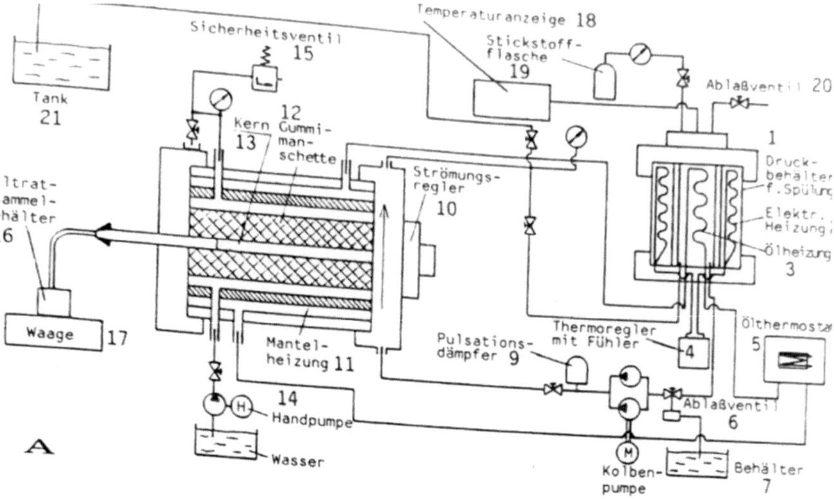

A

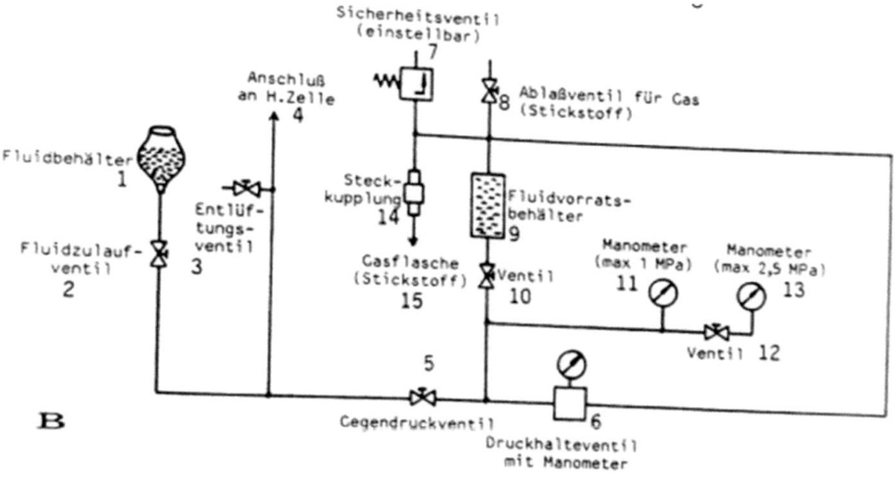

B

Encl. 24 A and B

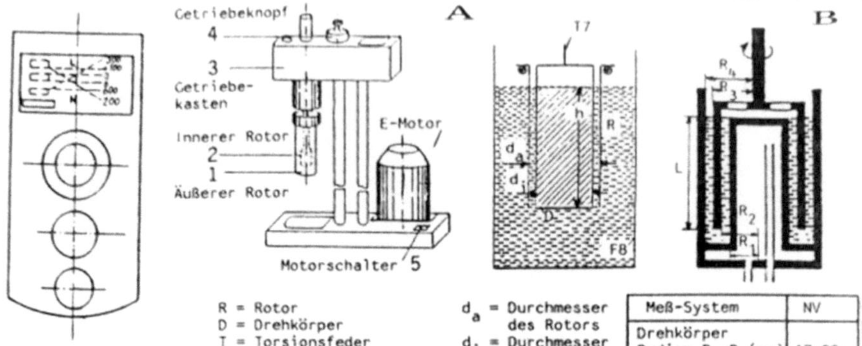

Meß-System	NV
Drehkörper Radius R_2;R_3(mm)	17,85; 20,1
Höhe L (mm)	60
Meßbecher Radius R1;R_4(mm)	17,5; 20,5

Encl. 25 A and B:

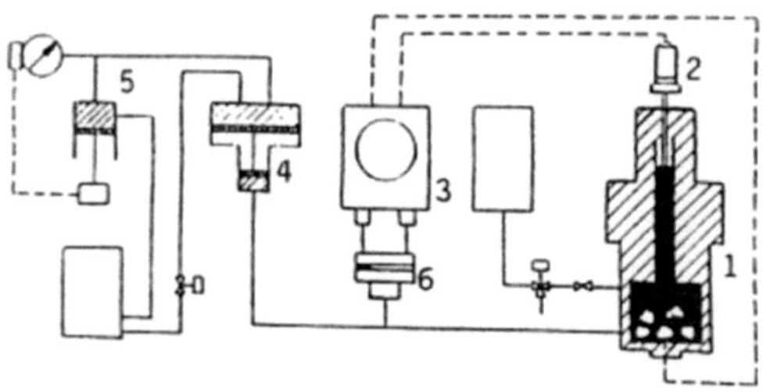

1. Autoklav
2. Motor für die Abtastspindel
3. Schreiber
4. Druckverstärker
5. Ölpumpe
6. Automatisches Abschaltventil

Encl. 26

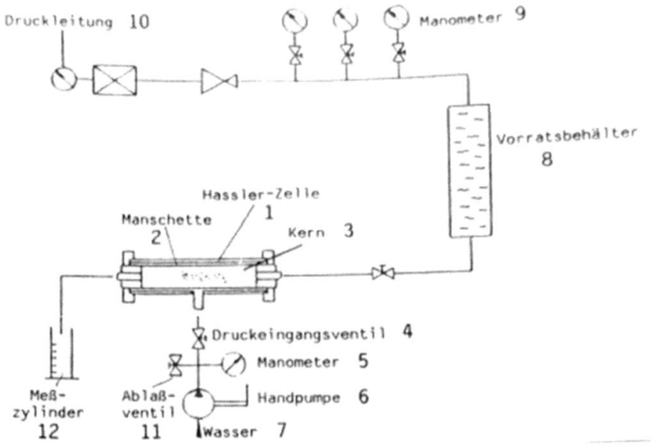

Encl. 27

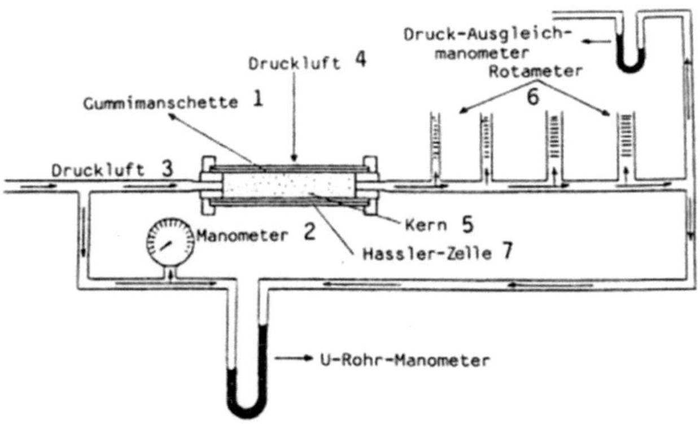

Encl. 28

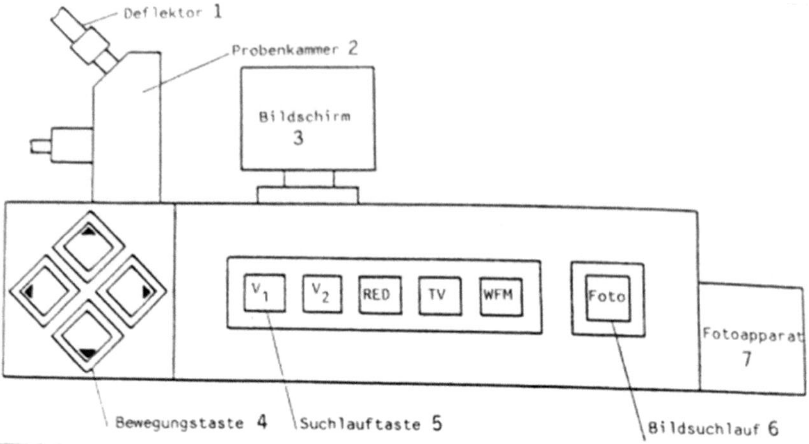

Encl. 29

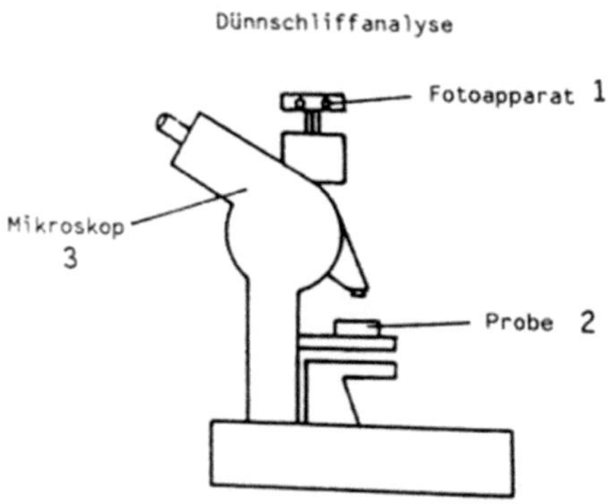

Encl. 30

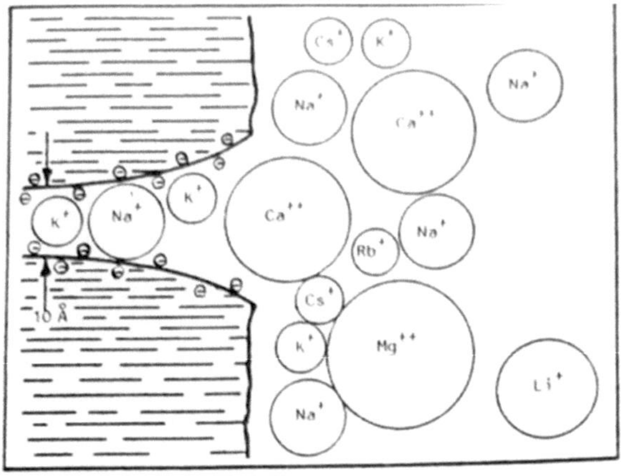

Encl. 31

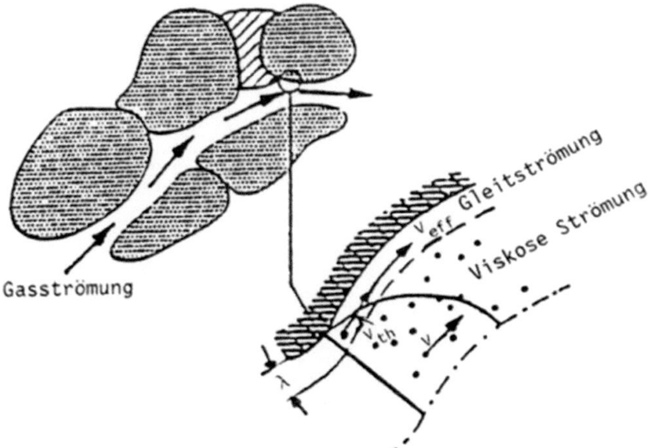

Encl. 32

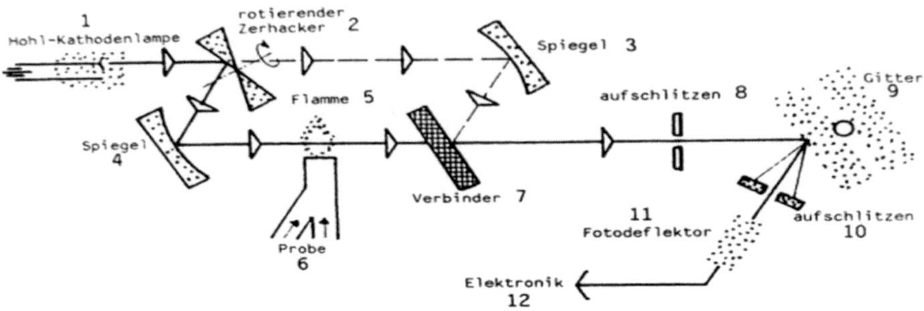

Encl. 33

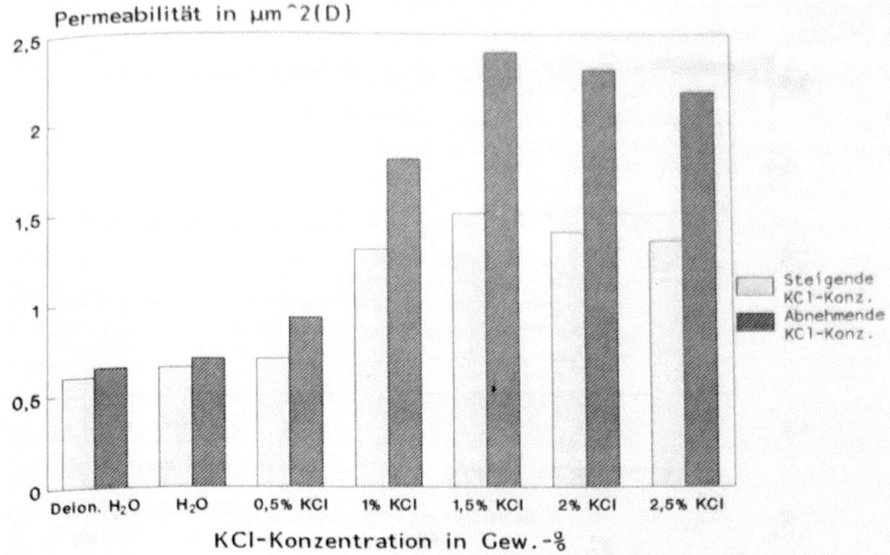

Encl. 34

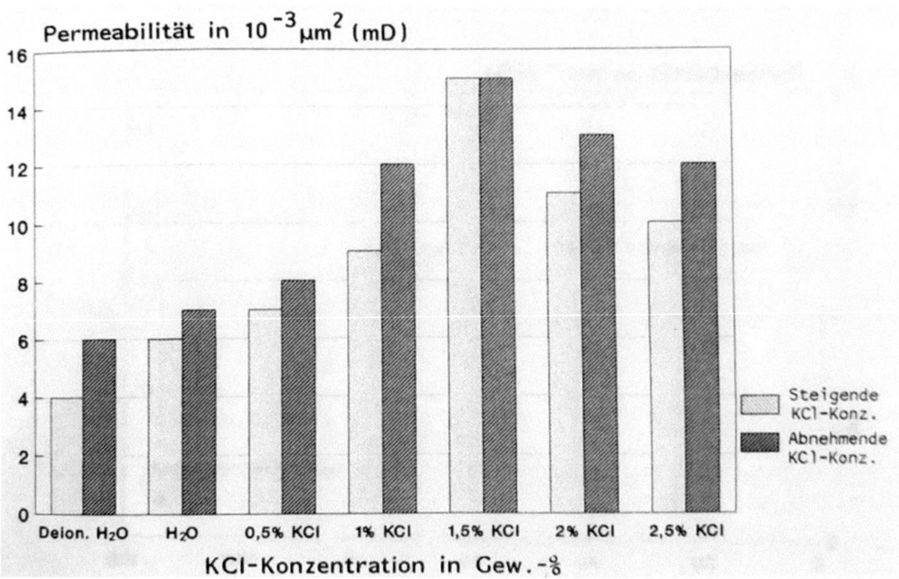

Encl. 35

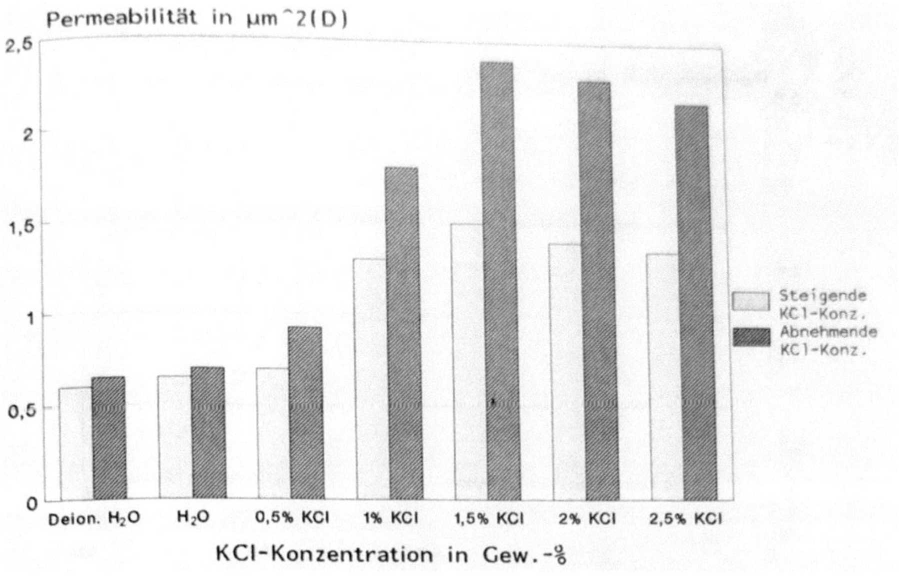

Encl. 36

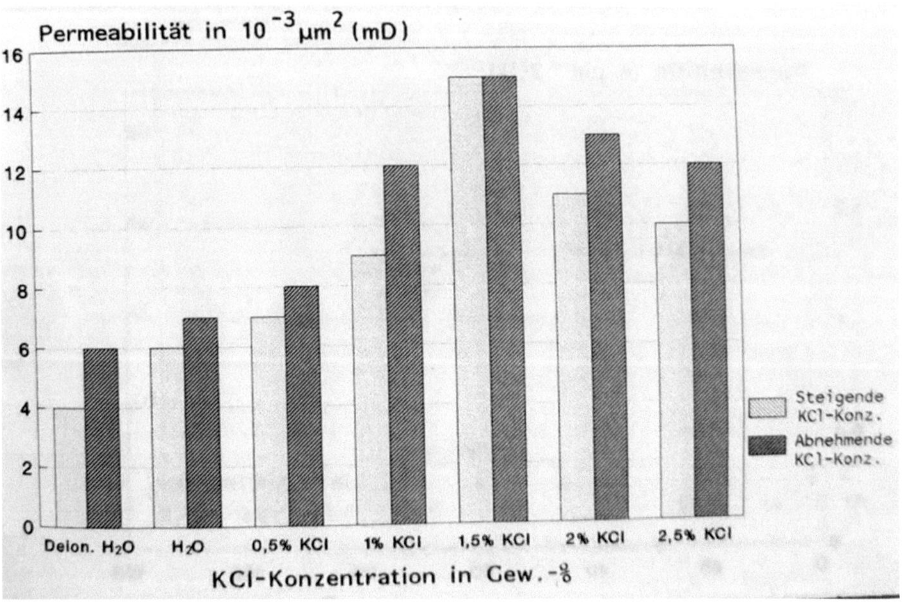

Encl. 37

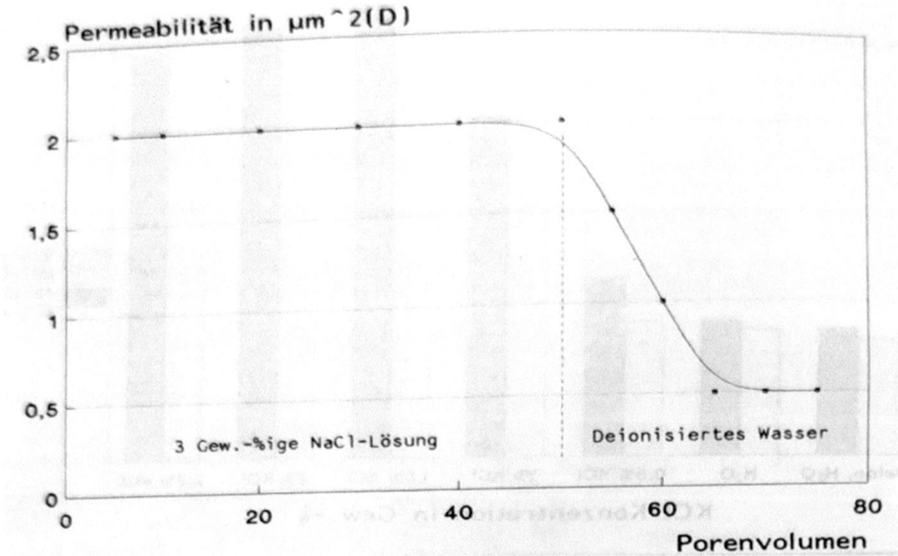

Encl. 38

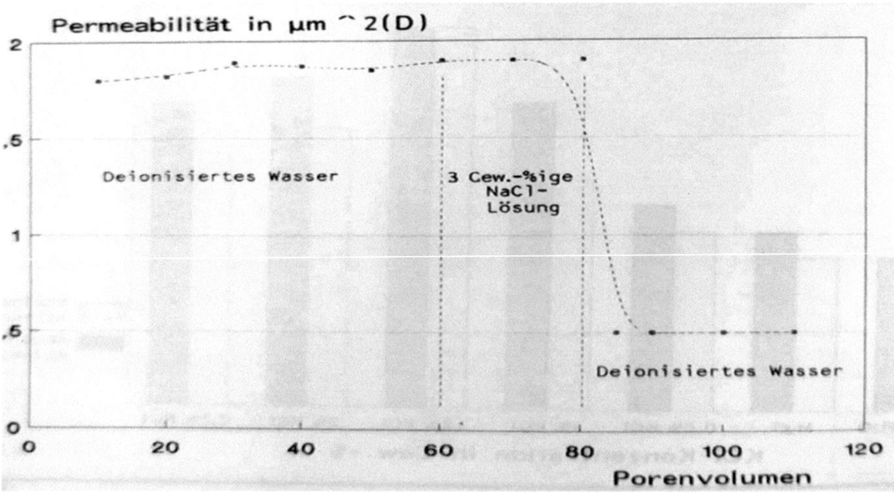

Encl. 39

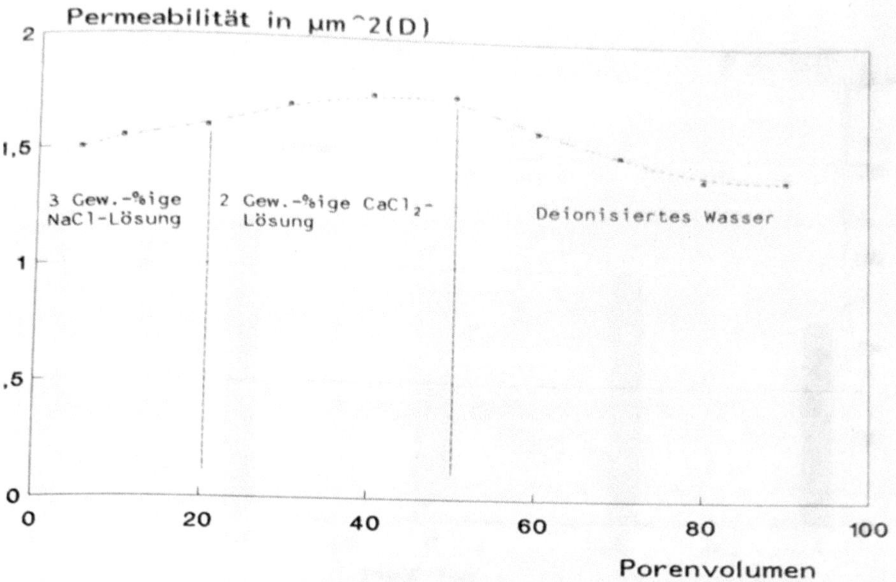

Encl. 40

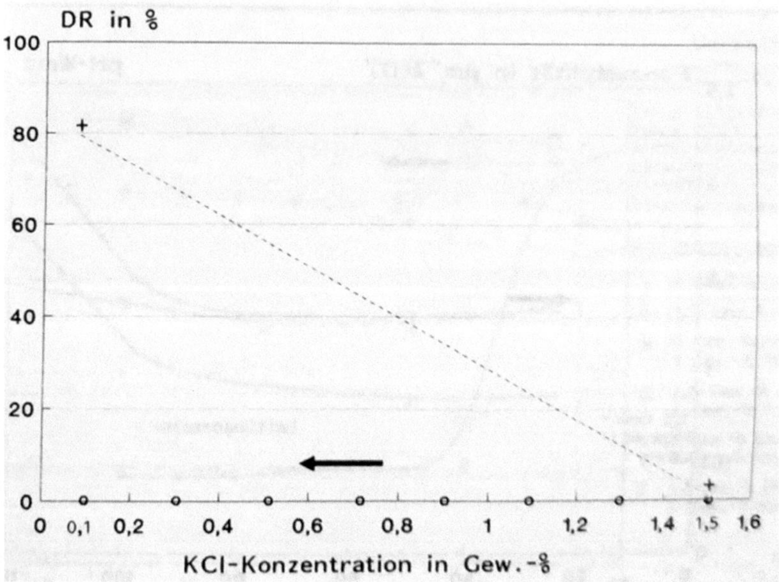

Encl. 41

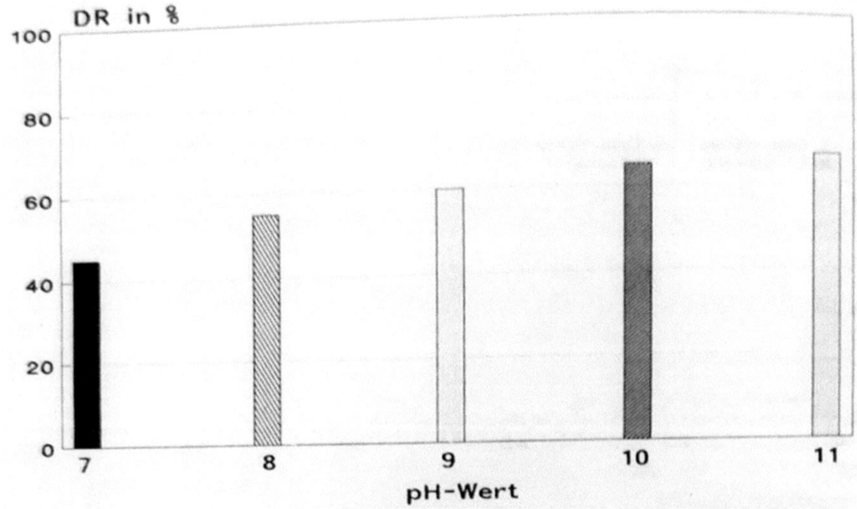

Encl. 42

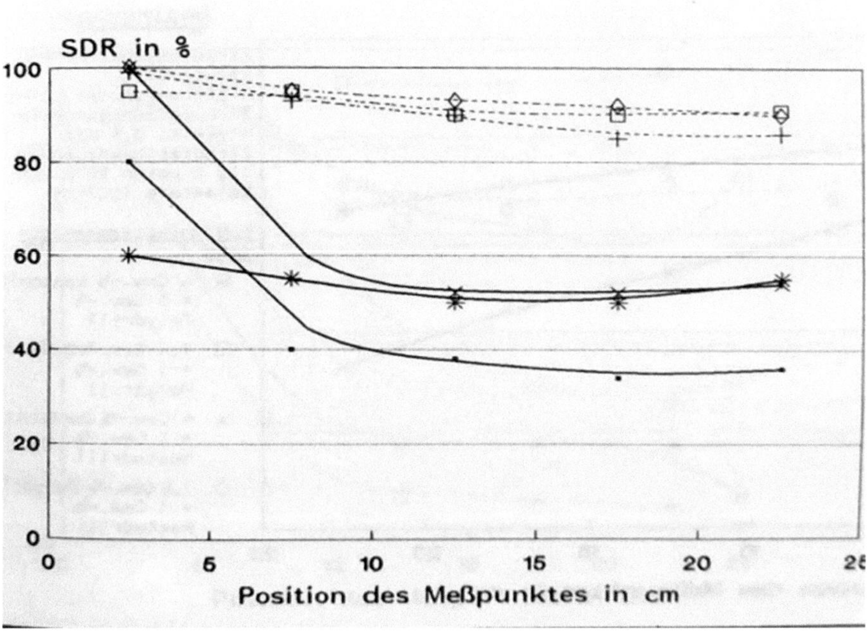

Encl. 43

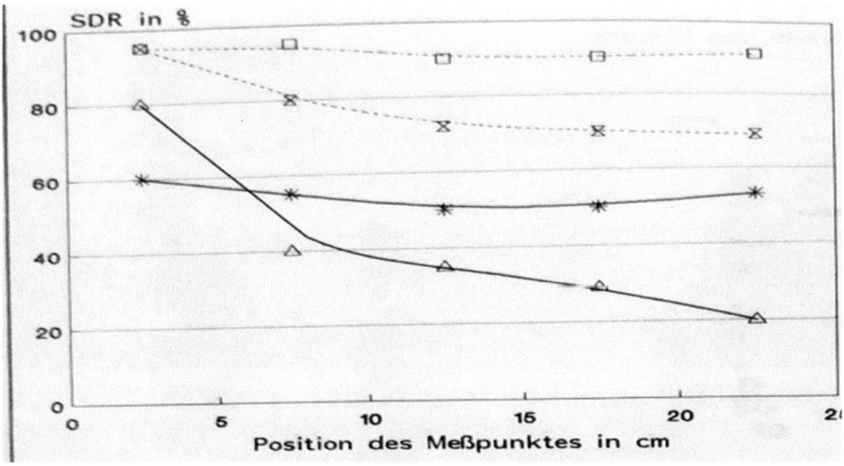

Encl. 44

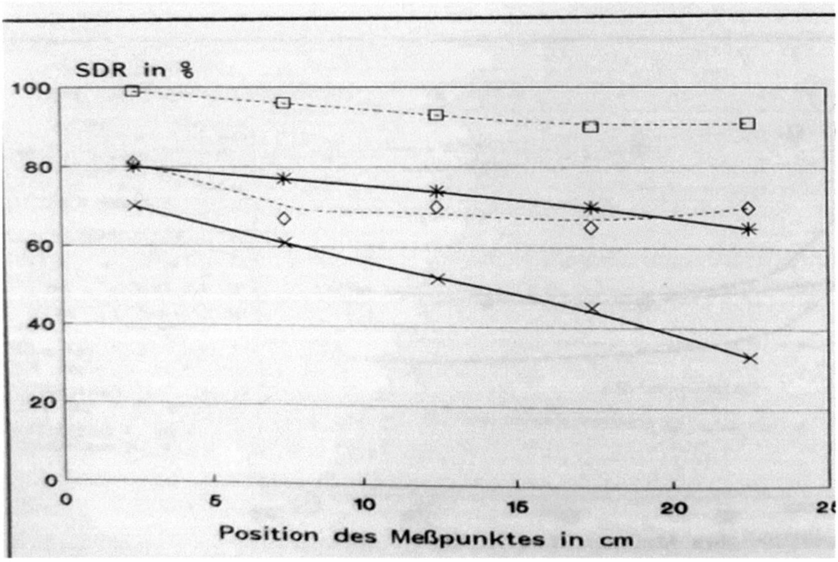

Encl. 45

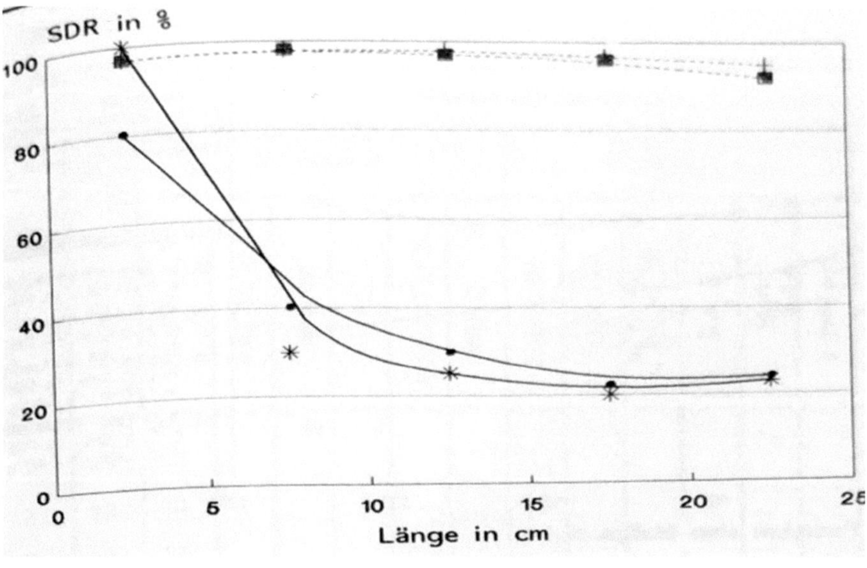

Encl. 46

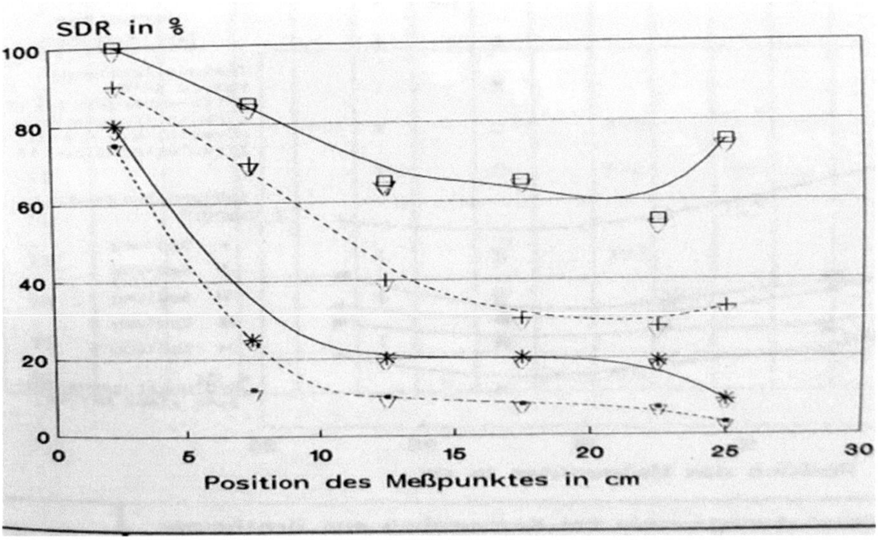

Encl. 47

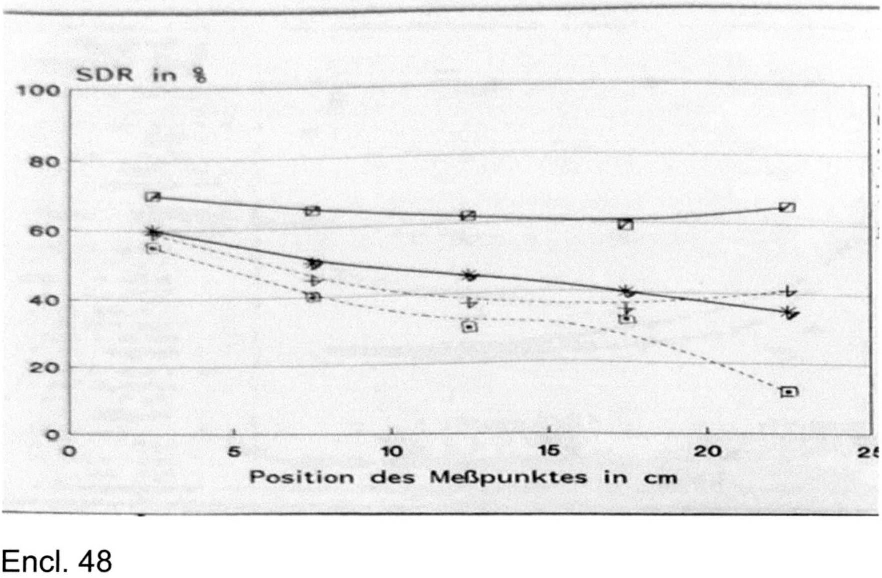

Encl. 48

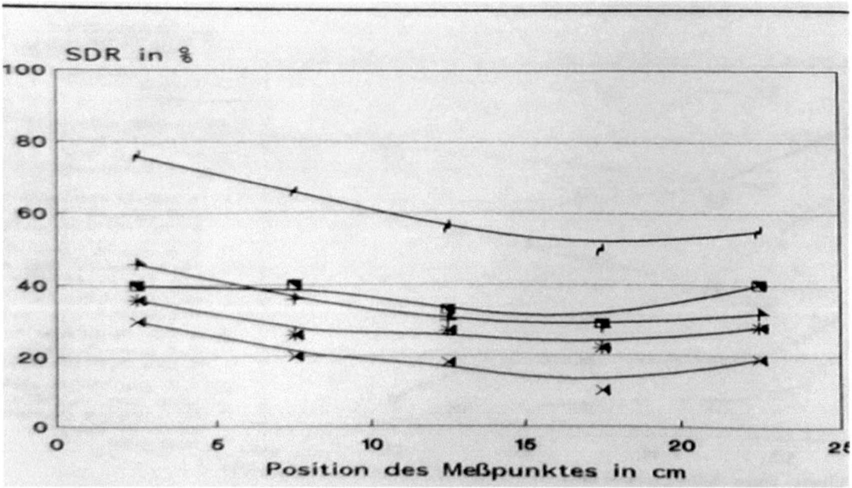
SDR in %
100
80
60
40
20
0
0
5
10
15
20
25
Position des Meßpunktes in cm

Encl. 49

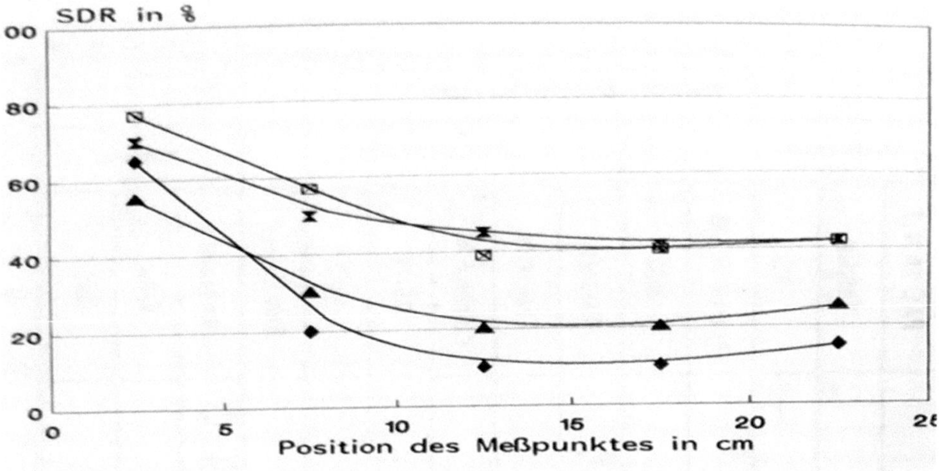

Encl. 50

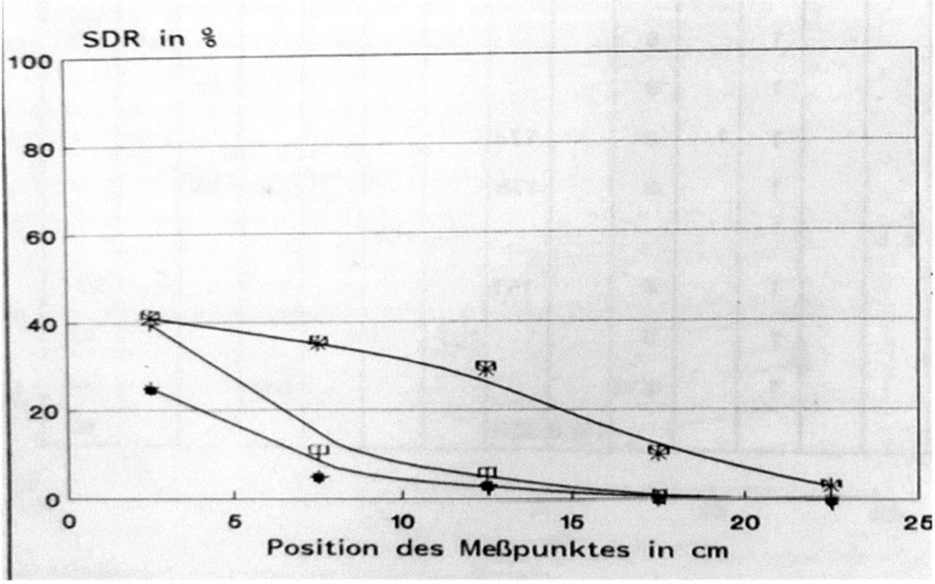

Encl. 51

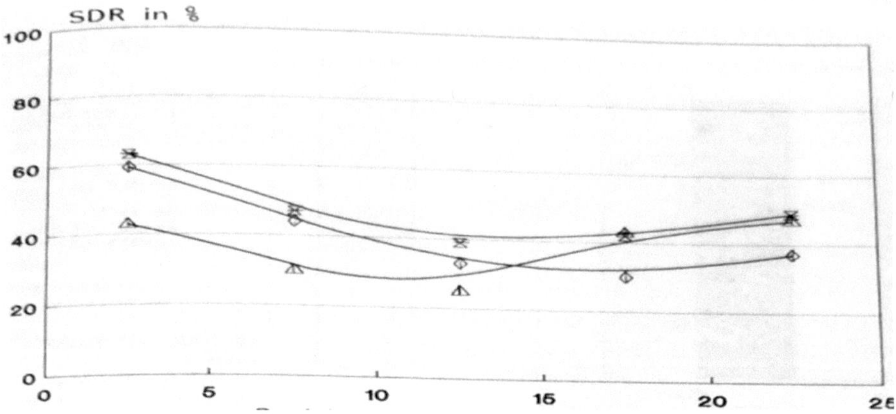

Encl. 52

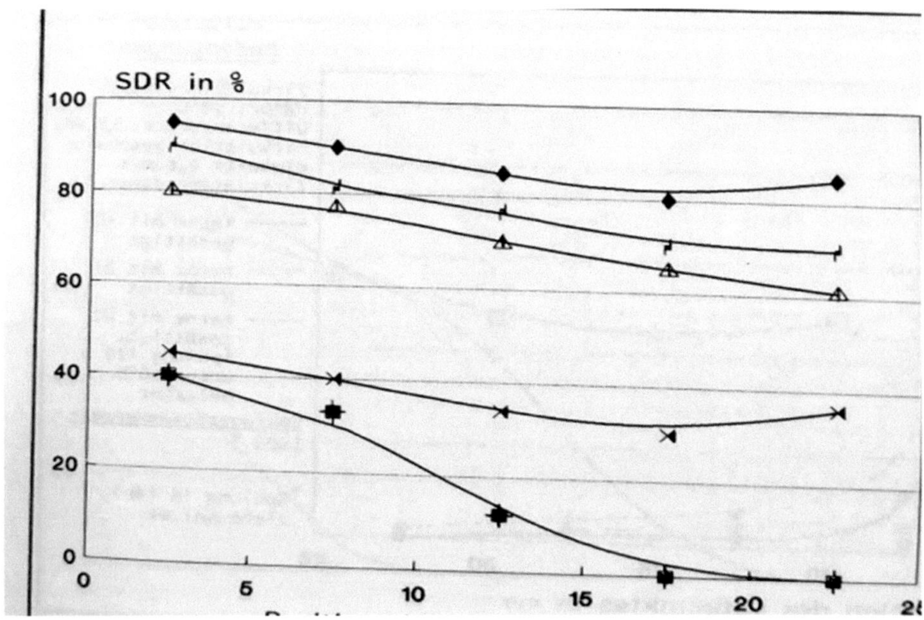

Encl. 53

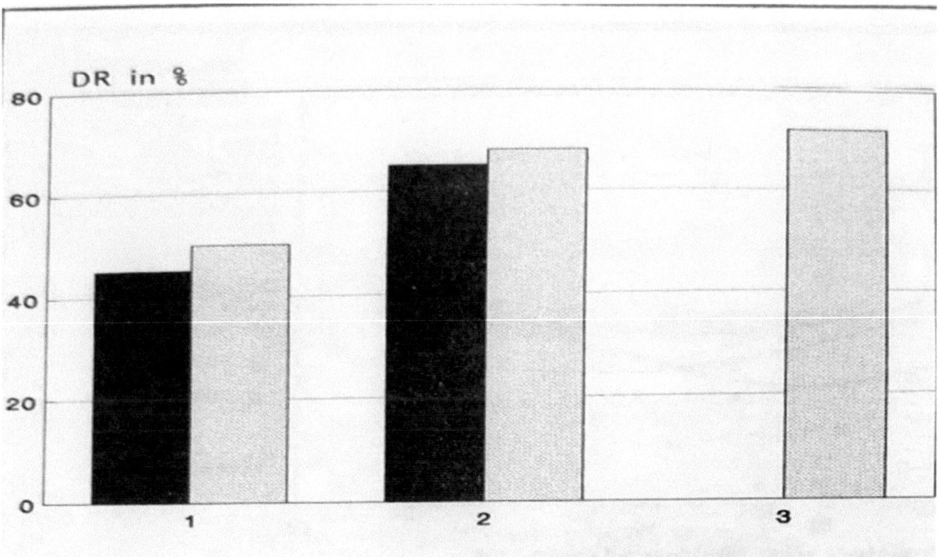

Encl. 53

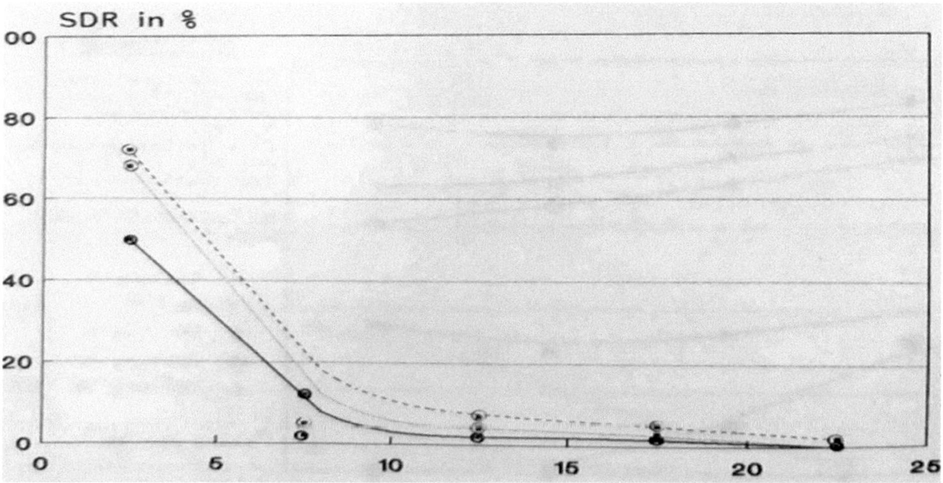

Encl. 54

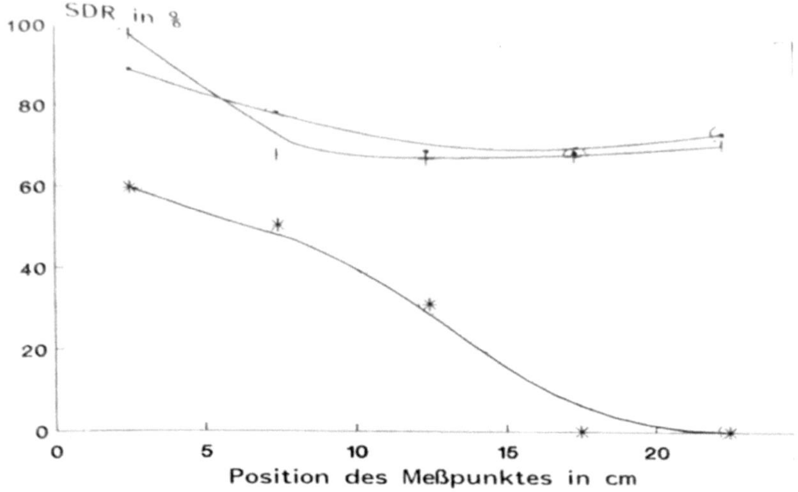

Encl. 55

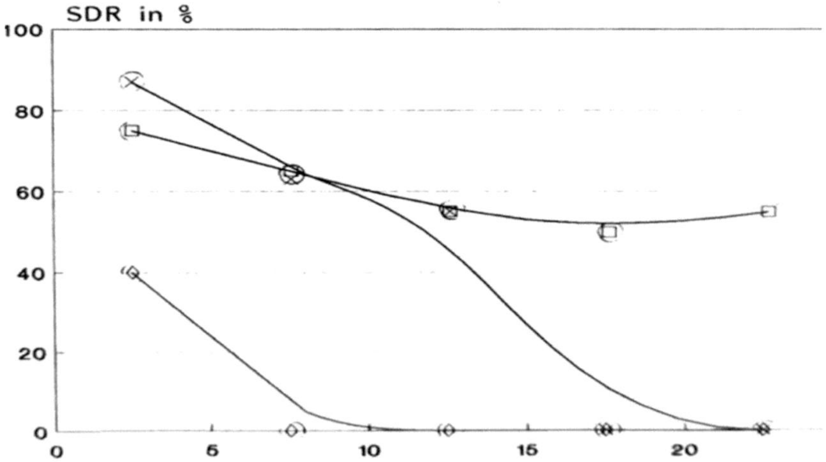

Encl. 56

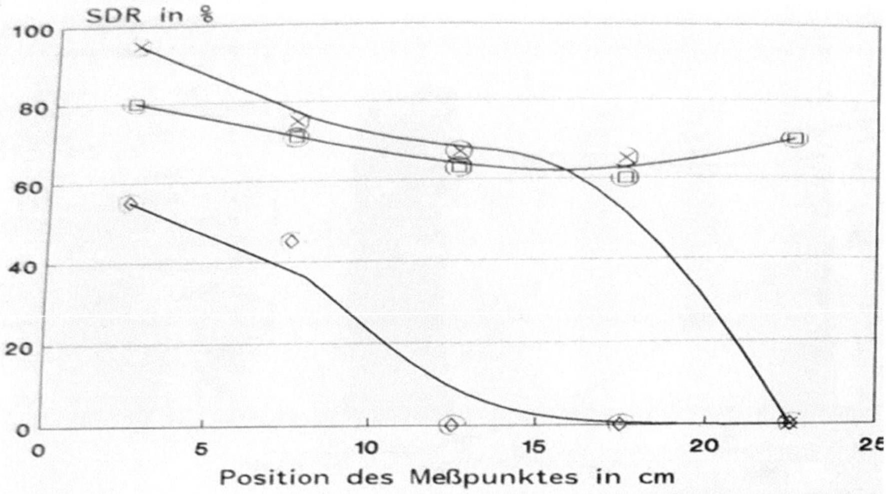

Encl. 57

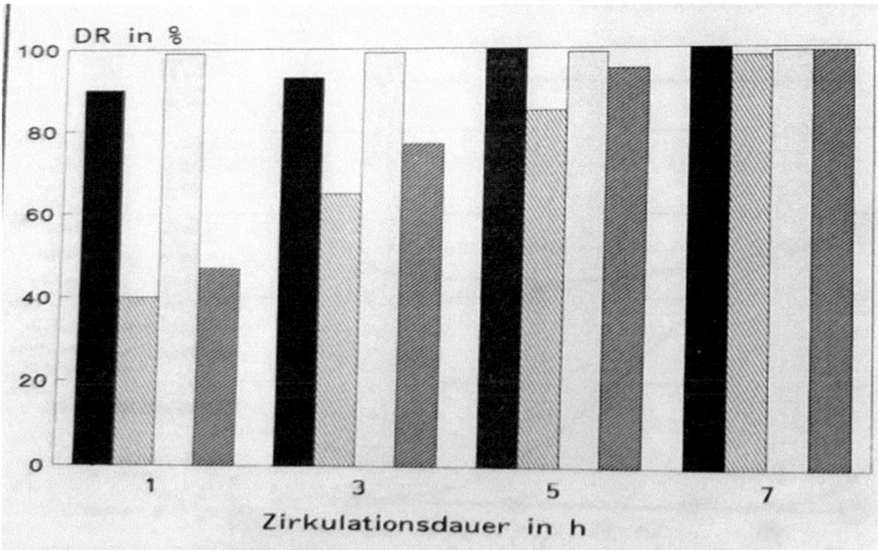

Encl. 58

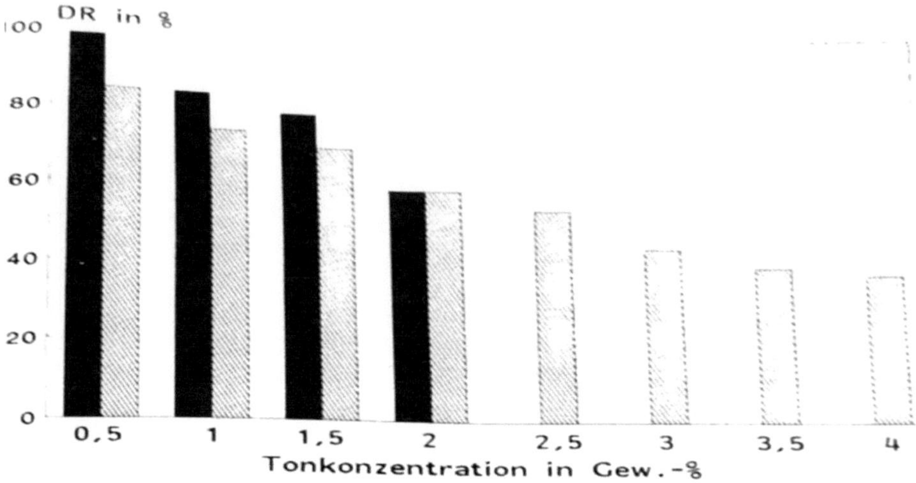

Encl. 59

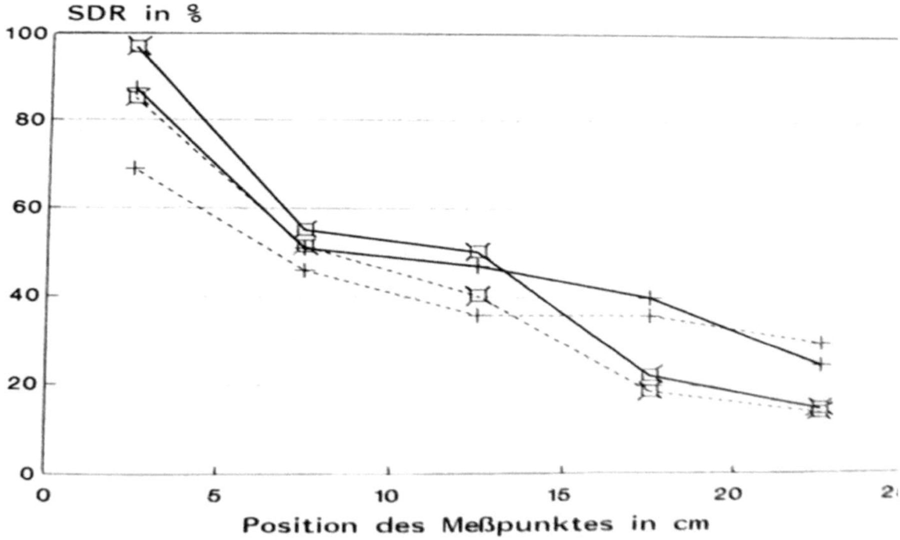

Encl. 60

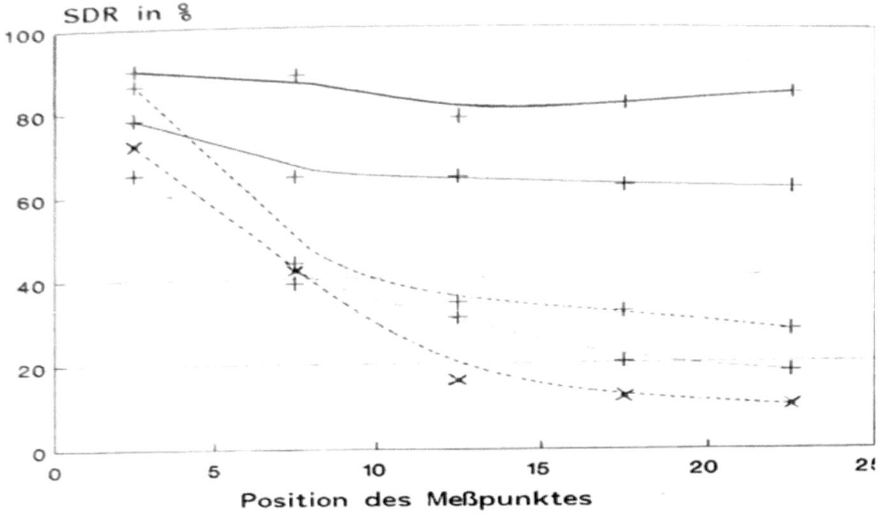

Encl. 61

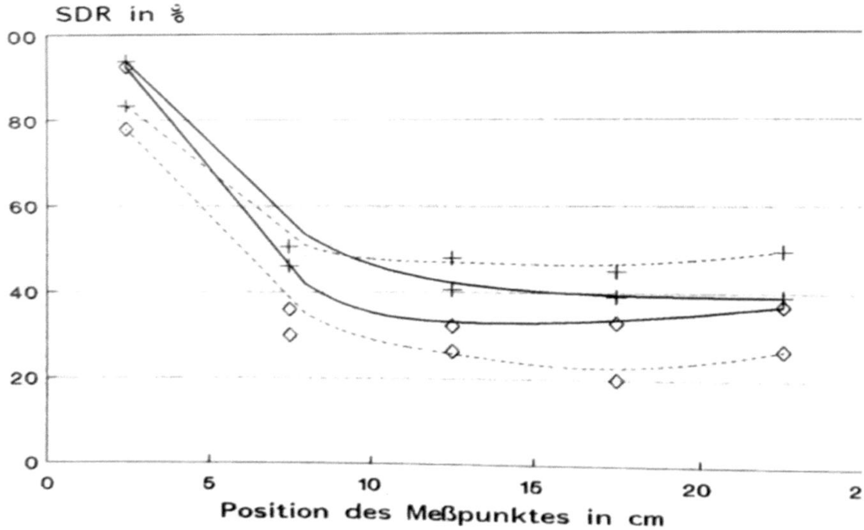

Encl. 62

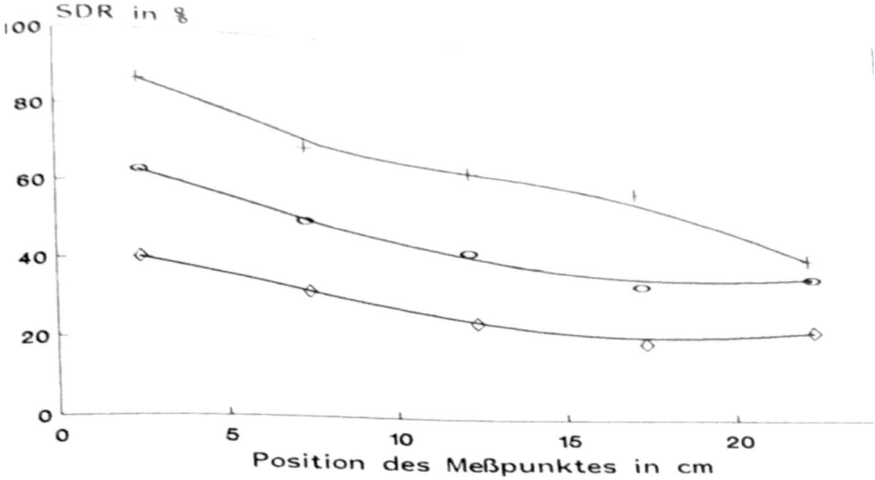

Encl. 63

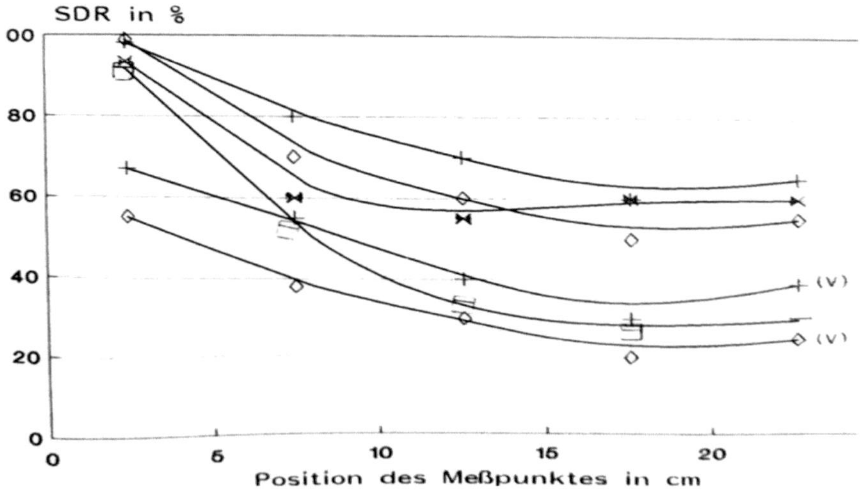

Encl. 64

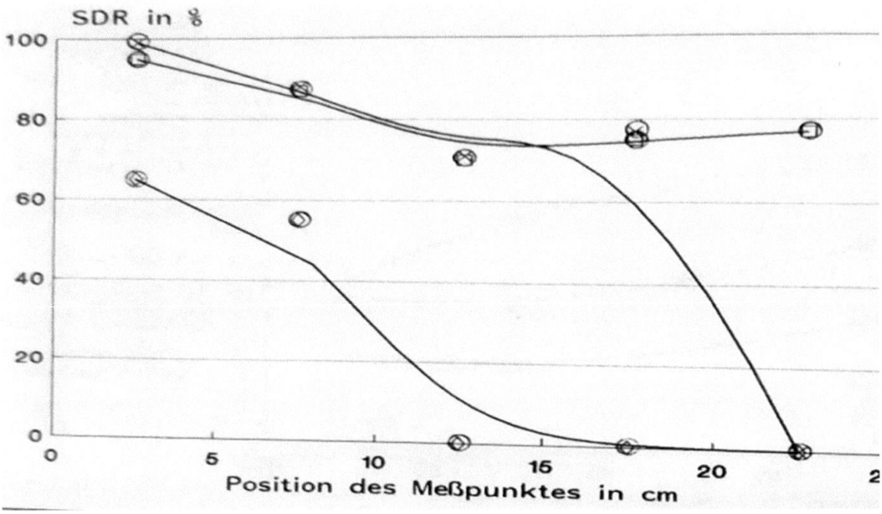

Encl. 65

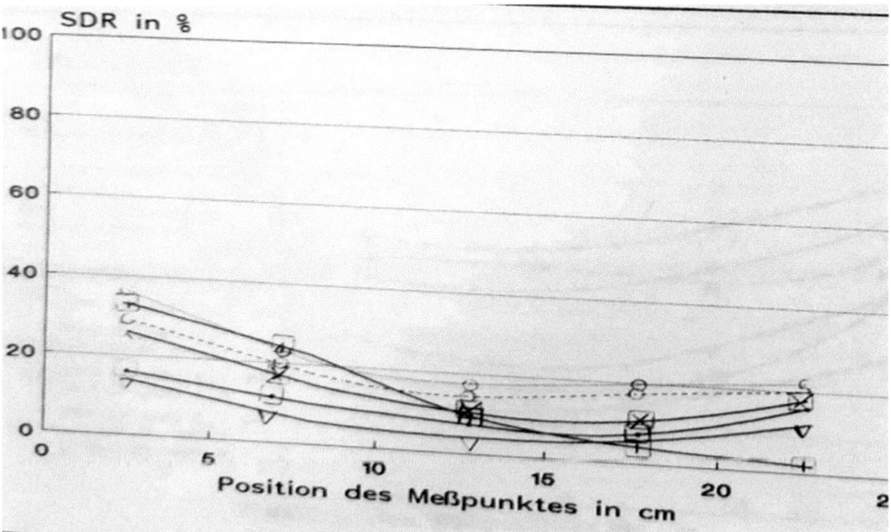

Encl. 66

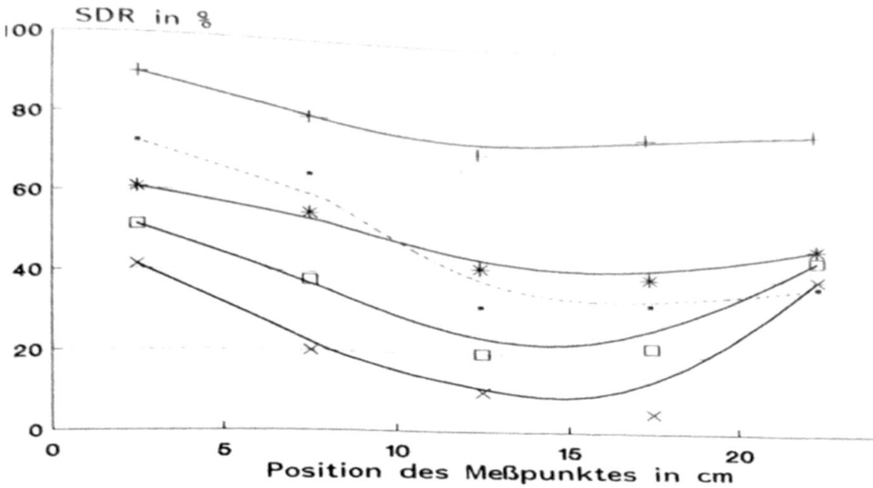

Encl. 67

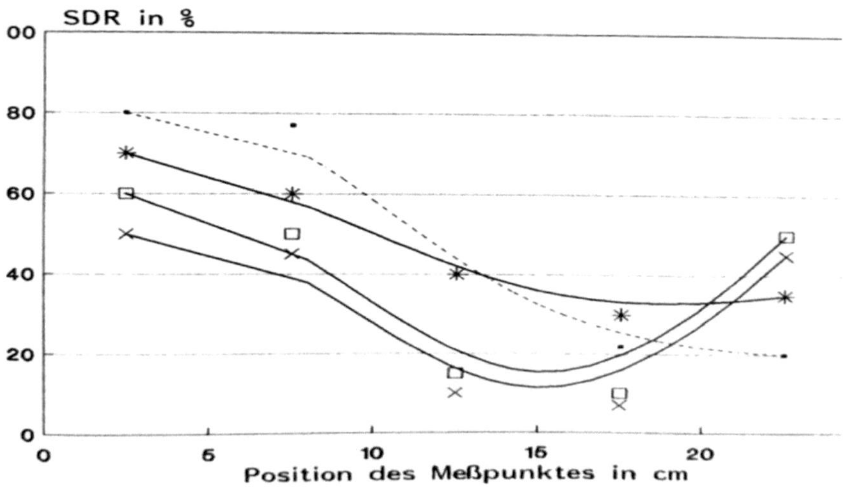

Encl. 68

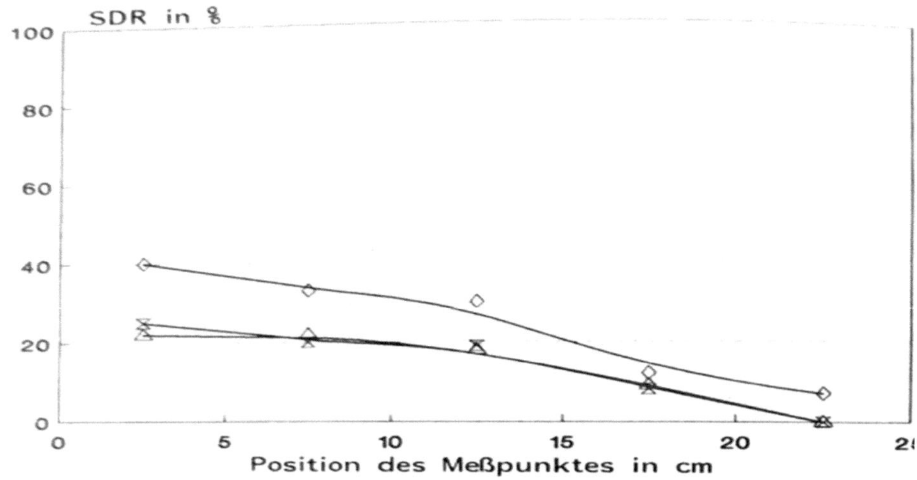

Encl. 69

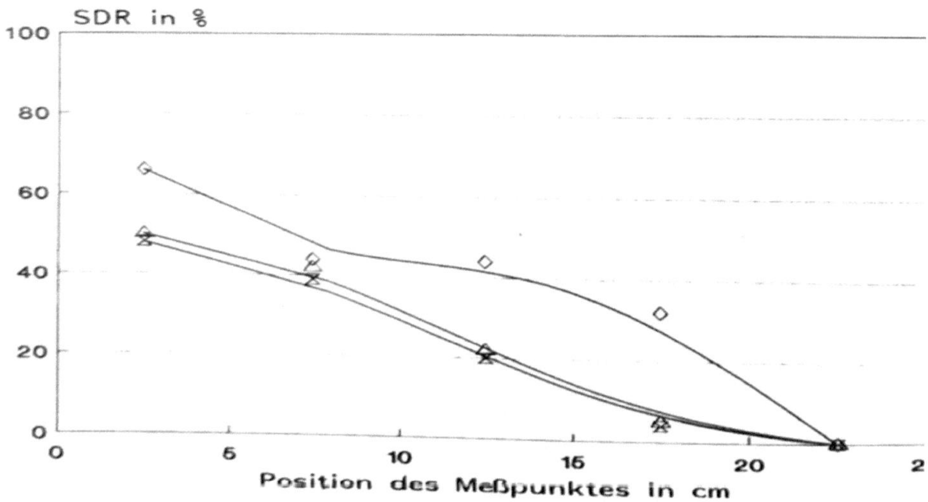

Encl. 70

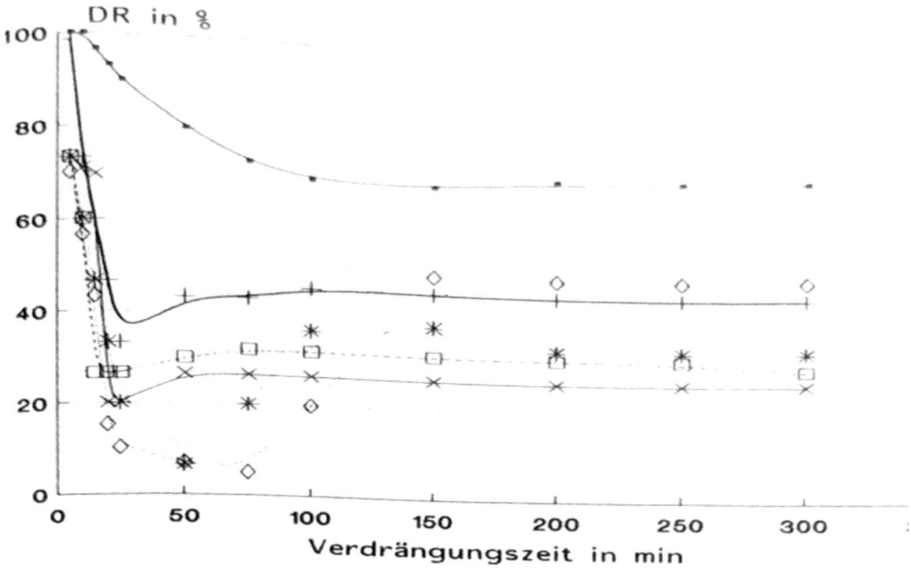

Encl. 71

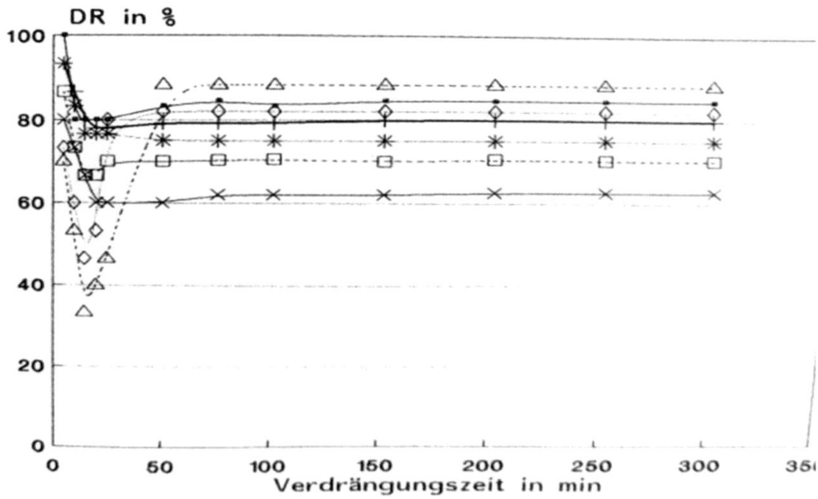

Encl. 72

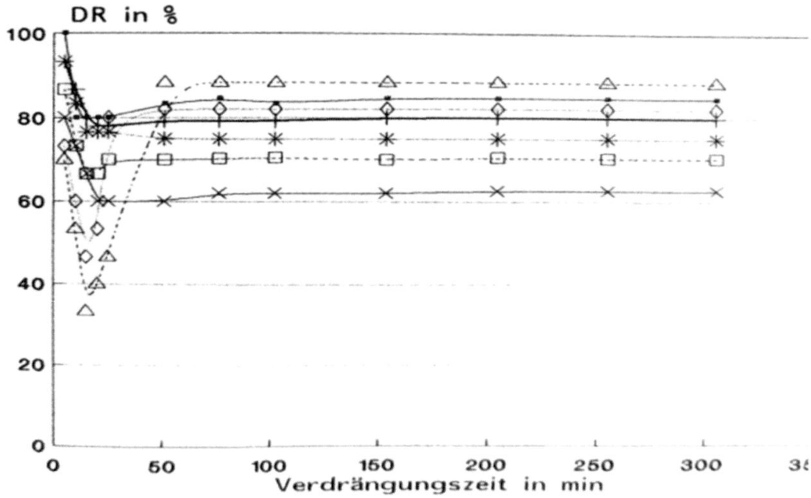

Encl. 73

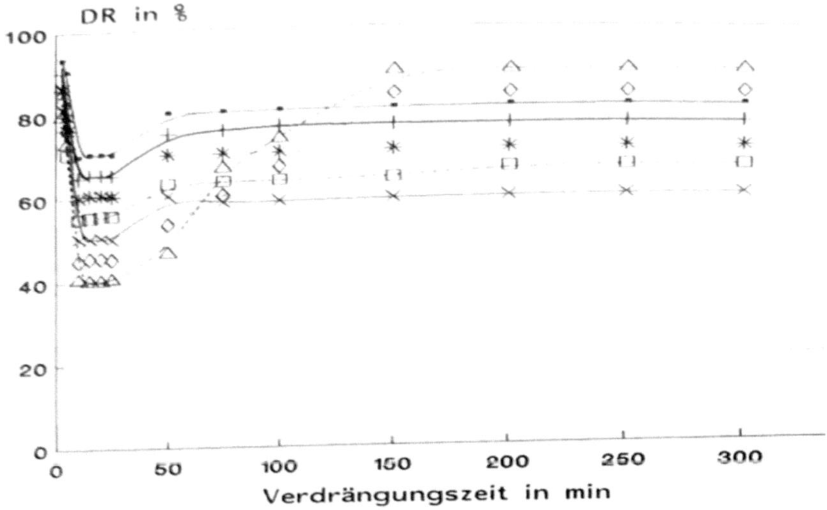

Encl. 74

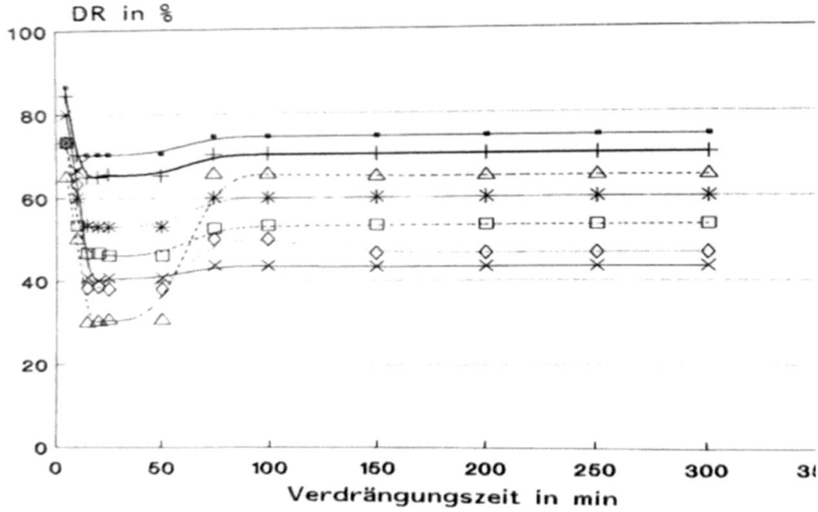

Encl. 75

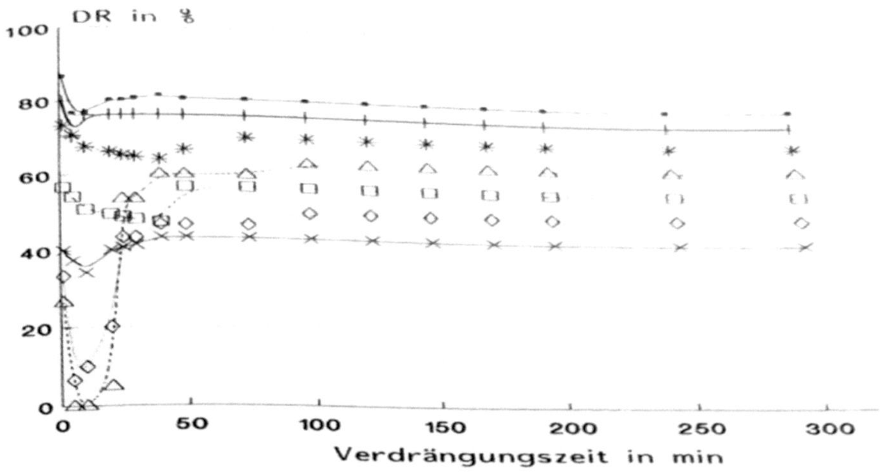

Encl. 76

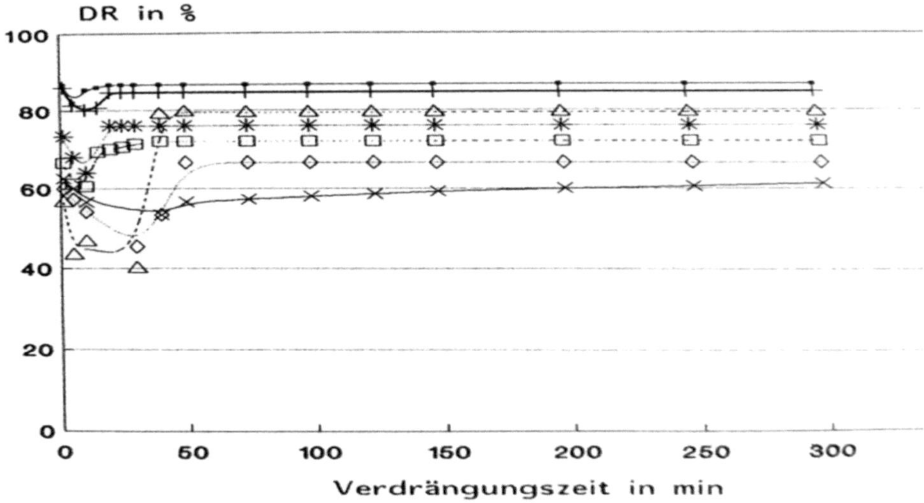

Encl. 77

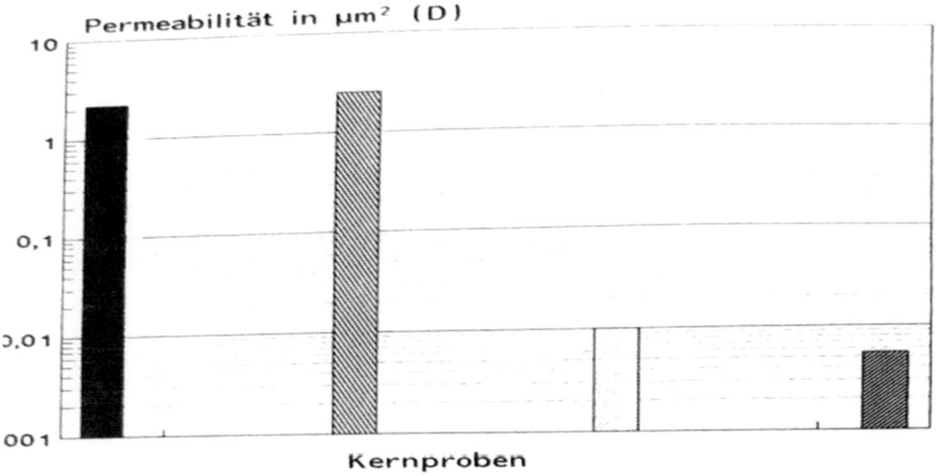

Encl. 78

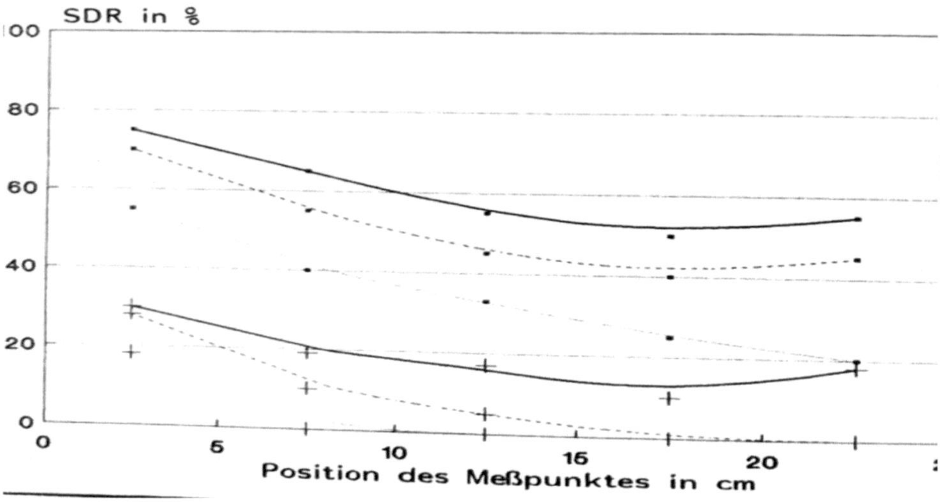

Encl. 79

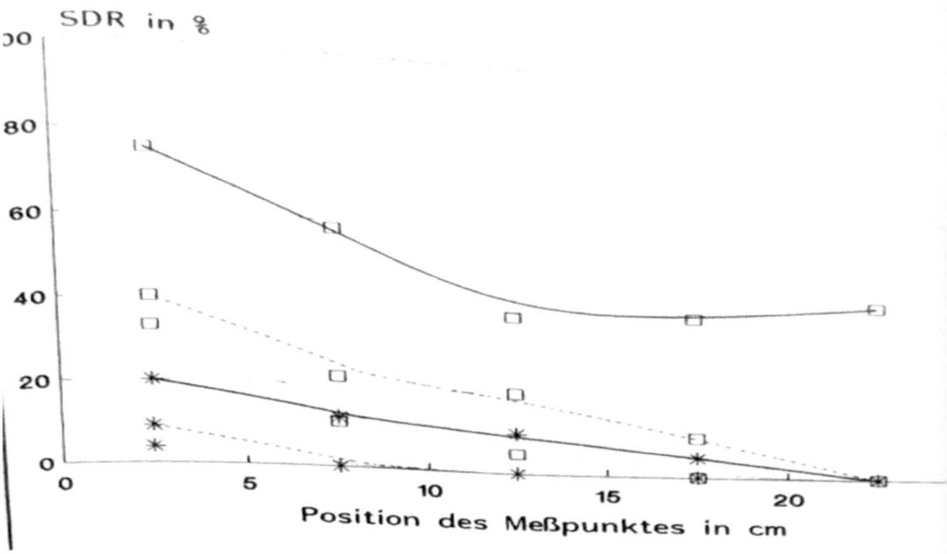

Encl. 81

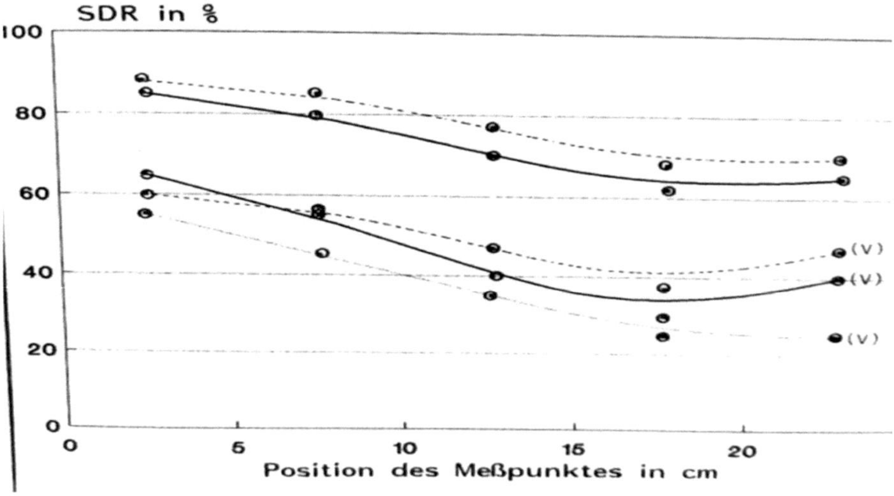

Encl. 82

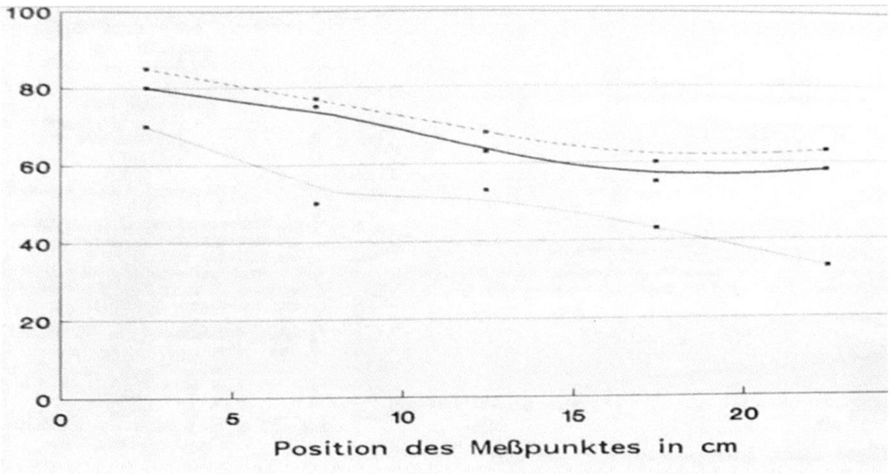

Encl. 83

Encl. 63

Encl. 64

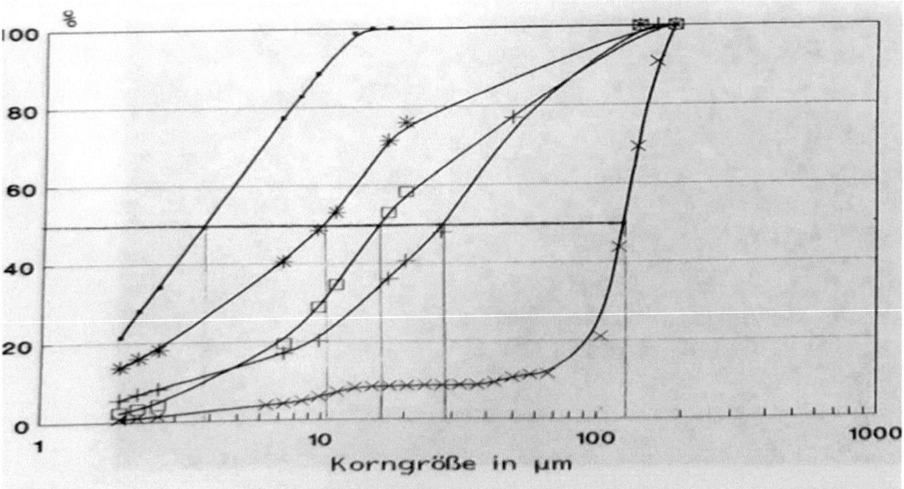

Encl. 65

The following pictures are not showing the influence of clay, polymers.

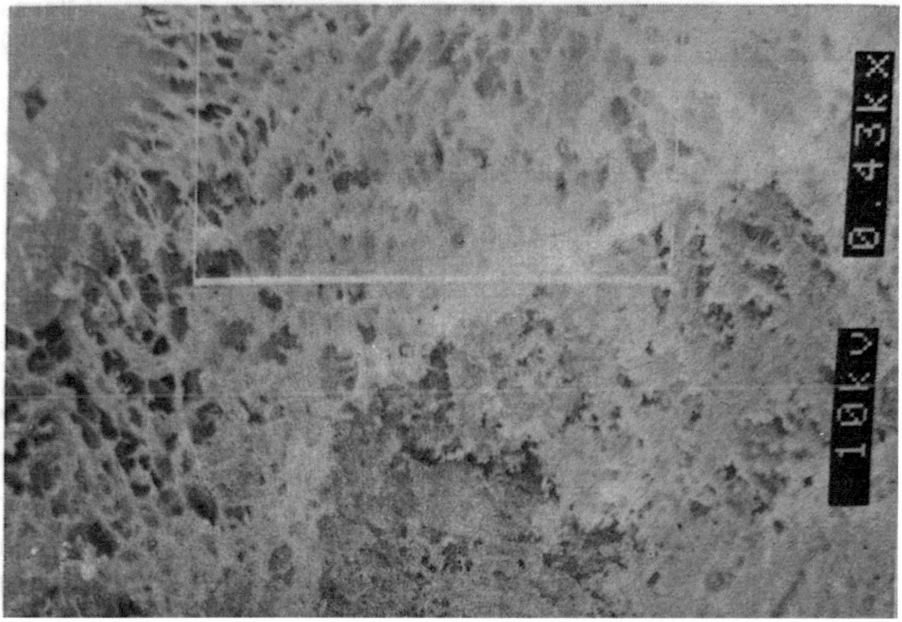

Encl. 44: Bentheimer Sandstone 4% Bentonite + 1% Polymer (first segment of core)

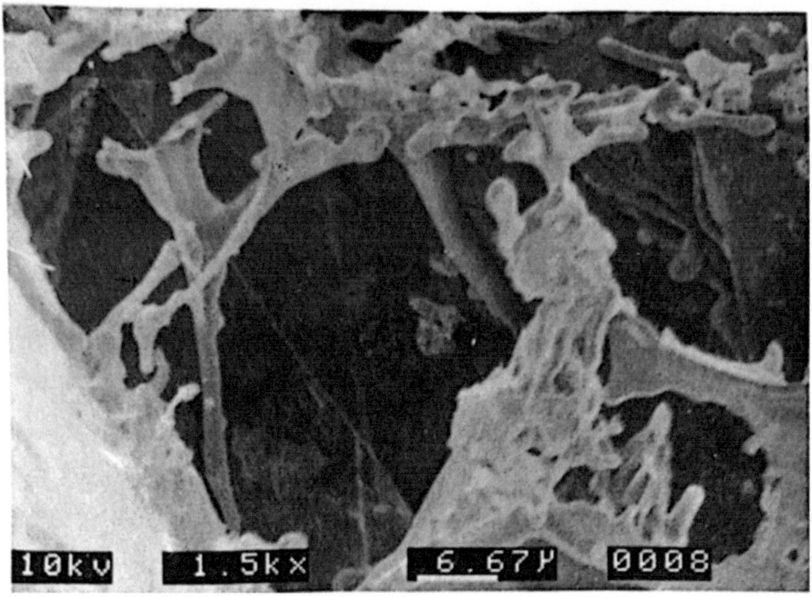

Encl. 45: Bentheimer Sandstone 4% Bentonite + 1% Polymer (fifth segment of core)

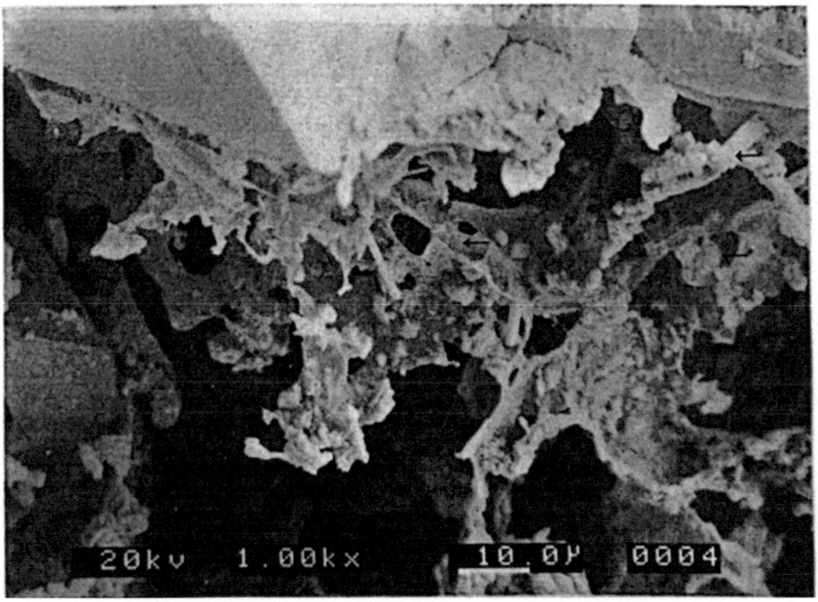

Encl. 46: Bentheimer Sandstone, fifth segment, DHT mud 1.5% + 1.5% Caco3

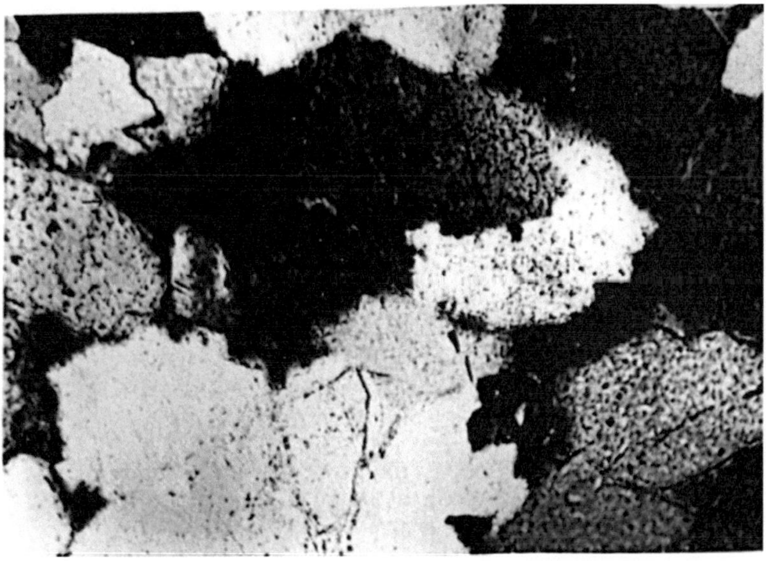

Encl. 47: Bentheimer Sandsone, 15% DHT, first segment

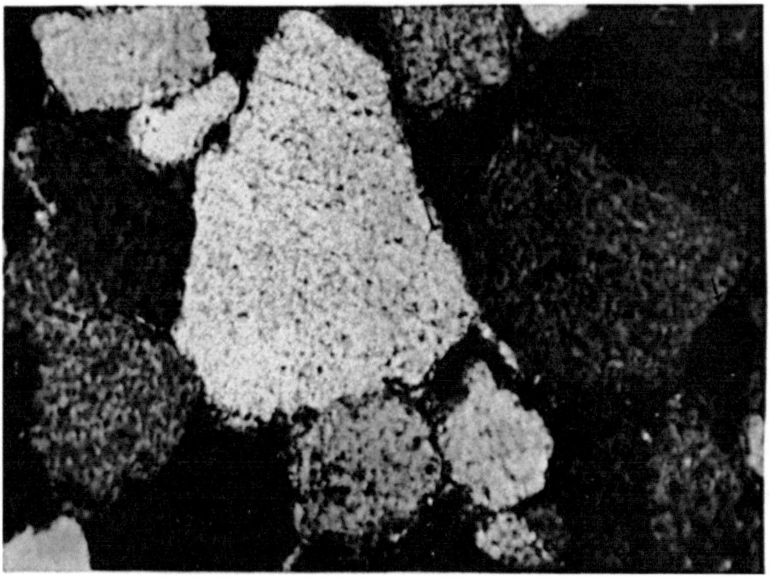

Encl. 48: Bentheimer Sandsone, 15% DHT, fifth segment

Encl. 49: Bentheimer Sandstone, first segment, 1.5 DHT + Polydrill + Cacl2

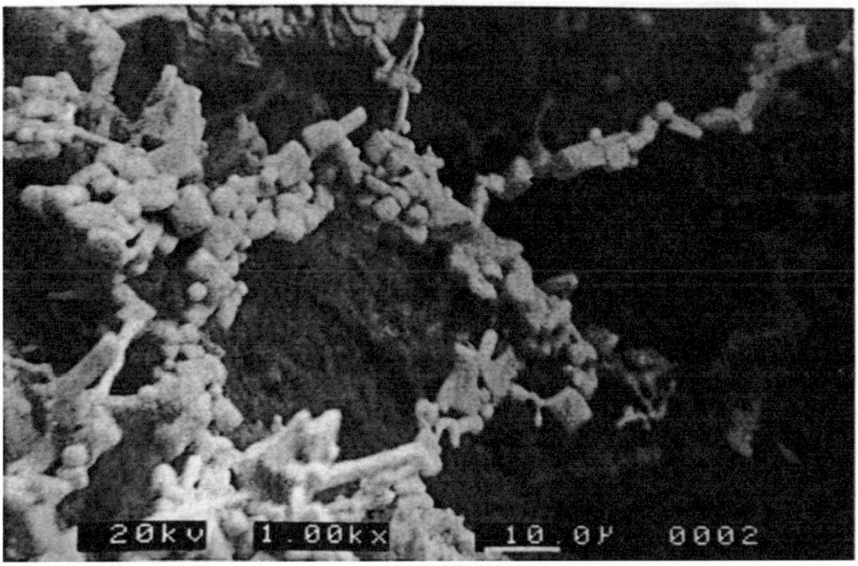

Encl. 50: Bentheimer Sandstone, fifth segment, 1.5 DHT + Polydrill + Cacl2

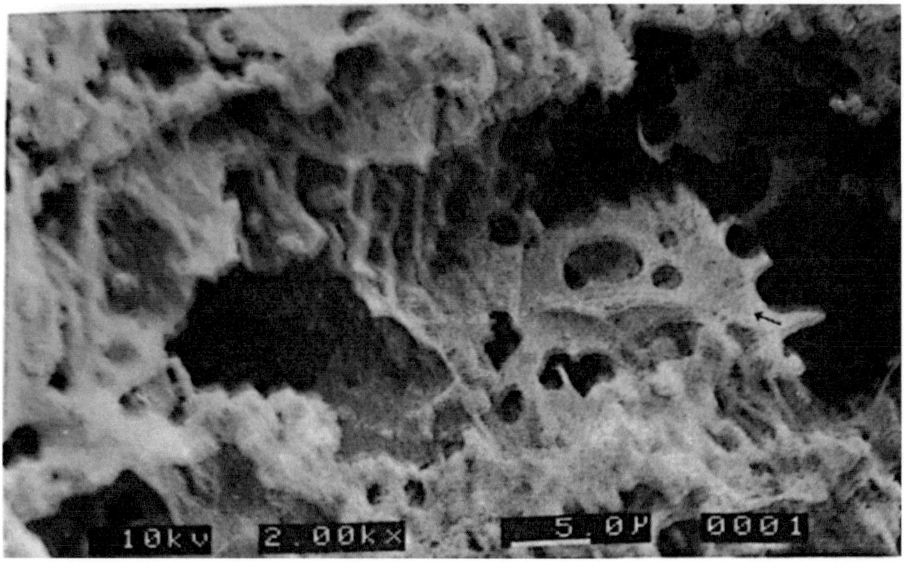

Encl. 51: Bentheimer Sandstone, first segment, 1.5% DHT + 10% Caco3

Encl. 52: Bentheimer Sandstone, fifth segment, 1.5% DHT + 10% Caco3

Encl. 53: Oberkichner Sandstone, first segment, 1.5% DHT

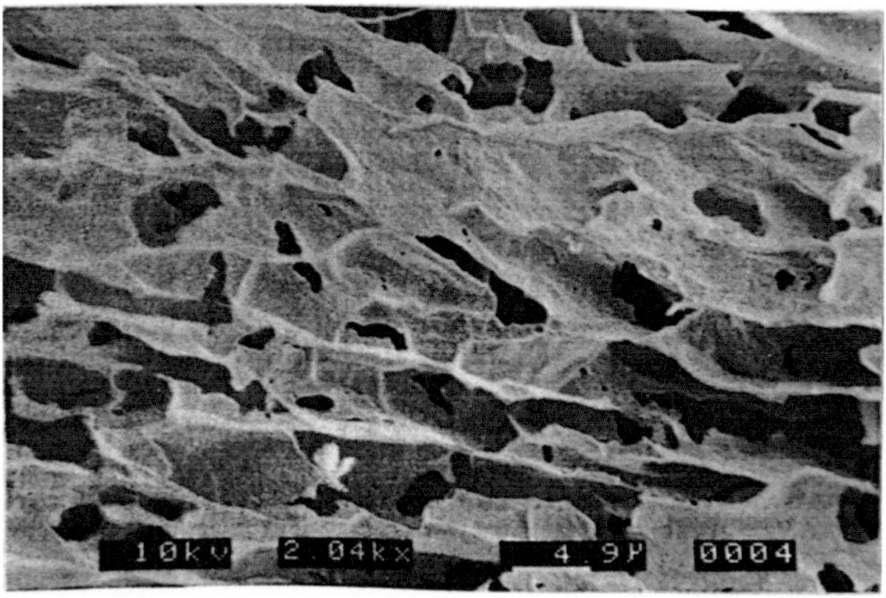

Encl. 54: Oberkichner Sandstone, first segment, 1.5% DHT, bigger picture

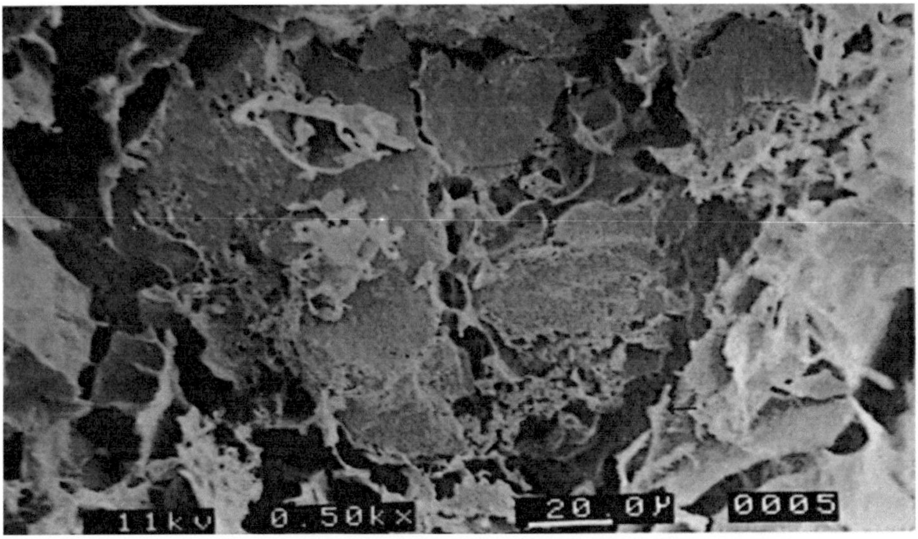

Encl. 55: Oberkichner Sandstone, fifth segment, 1.5% DHT, bigger picture

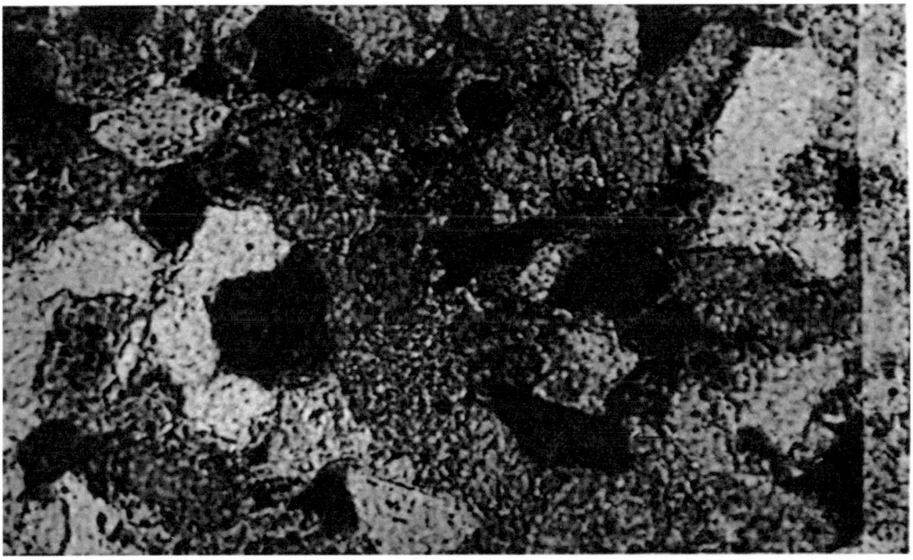

Encl. 56: Oberkichner Sandstone, first segment, 1.5% DHT

Encl. 57: Oberkichner Sandstone, fifth segment, 1.5% DHT

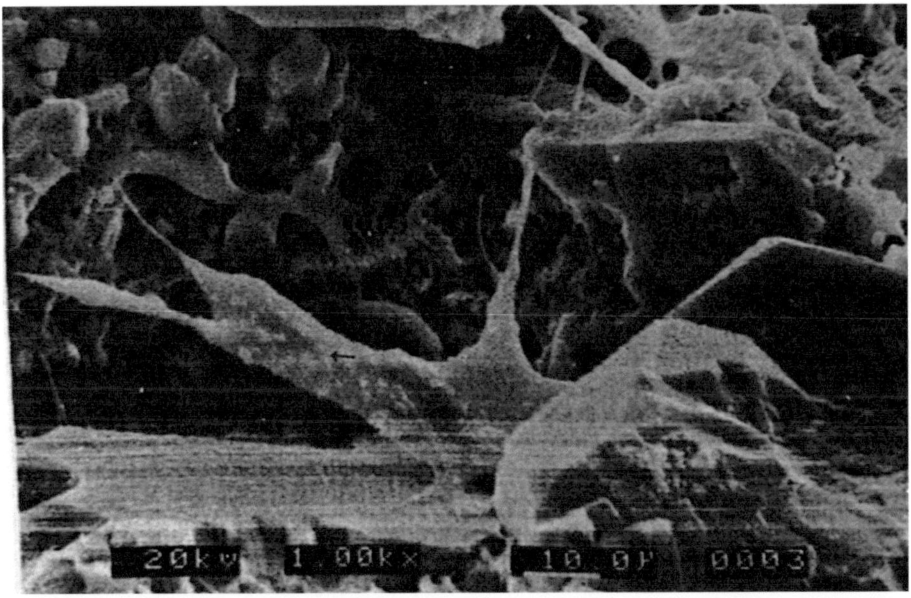

Encl. 58: Oberkichner Sanstone, first segment after formation damage with treatment with Hydrop[ac

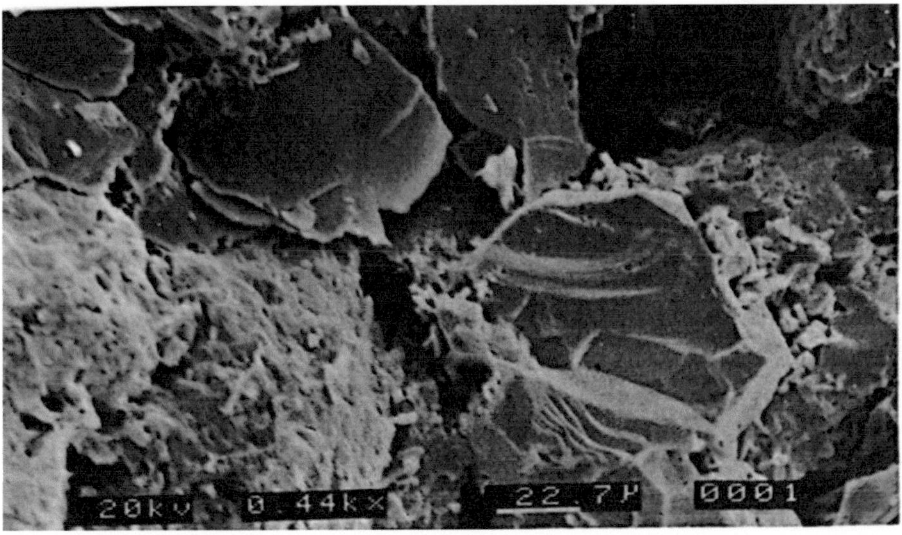

Encl. 59: Oberkichner Sandstone, first segment after treatment with Hydropac + 0.01 Am-SP

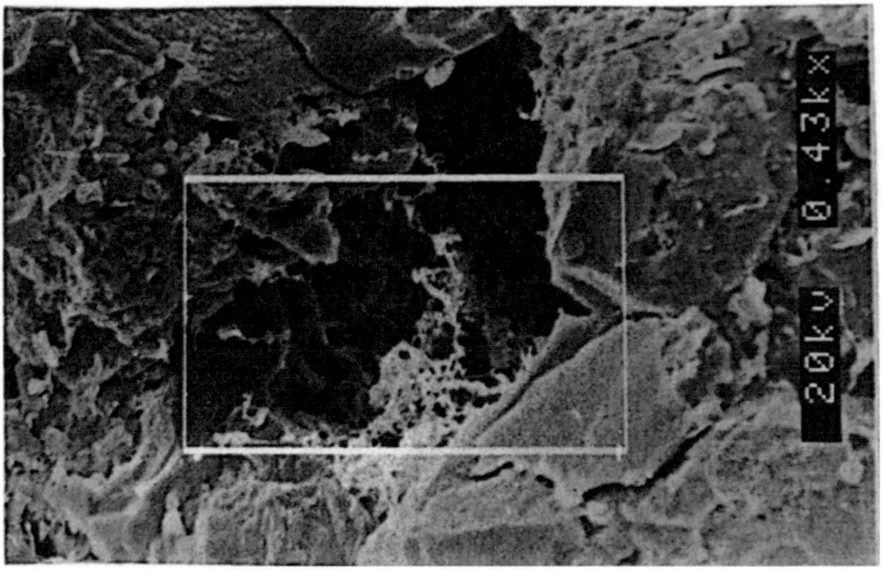

Encl. 60: Oberkichner Sandstone, fifth segment after treatment with Hydropac + 0.01 Am-SP

Encl. 61: REM – Caco3

Encl. 62: Bentheimer Sandstone

25. REFERENCES

JILLOGUE /1/ Ghofrani, R.; Delius, A. :	Aussagefähigkeit der API-Filtratwerte über das Filtrationsverhalten von Bohr- spülungen unter Bohrlochbedingungen. EEK-Z. 107 (1991), 9, S. 361-363
/2/ Ghofrani, R.:	Untersuchungen der dynamischen Filtration der im Bohrbetrieb eingesetzten wasser-basischen Bentonitspülungen und der Möglichkeit einer Abtragung des Filterkuchens mit Hilfe von Waschflüssigkeiten unter simulierten Bohrlochbedingungen. Dissertation, ITE, TU Clausthal 1975
/3/ Allen, D.;	Invasion Revisited Oilfield Review, July 1991, S. 10
/4/ Gray, G.R. ; Darly, H.C.H.; Rogers, W.F.	Composition and Properties of Oil Well Drilling Fluids, 4. Ed., Gulf Publishing Company, Houston, London, Paris, Tokyo 1980, S. 477
/5/ Chilingarian, V.; Vorabutr, P.	Drilling Fluids, 4. Ed., Gulf Publishing Company, Houston, London, Paris, Tokyo 1980, S. 477 Drilling and Drilling Fluids, 2. Ed., : Elsevier, Amsterdam, Oxford, New York 1984, S.331
/6/ Nowak, T.J.; Krueger, R.F.:	The Effect of Mud Filtrates and Mud Particles upon the Permeability of Cores Drill. & Prod. Prac., API 1951, S. 164-181
/7/ Ghofrani, R.:	Spülungstechnik - Skriptum zur Vorlesung - ITE, TU Clausthal, Oktober 1985
/8/ Muecke, T.W.:	Formation Fines and Factors Controlling Their Movement in Porous Media. JPT, Feb. 1979, S. 144-150
/9/ Stanislav, J.F.;	Pressure Transient Analysis Handbook, 1. Ed., Prentice Hall Englewood

cliffs, New Jersey 1990, S.34-59

/10/ Kuchuk, F.J.; Kirwan, P.A.: New Skin and Wellbore Storage Type Curves for Partially Penetrated Wells. SPE Formation Evaluation, Dec. 1987, S. 546-554

/11/ Ohen, H.A.; Civan, F. : Predicting Skin Effects Due to Formation Damage by Fines Migration. SPE 21675, 1991
Scanned with CamScanner

/12/ Muskat, M.: Physical Principales of oil Production. McGraw-Hill Book Company, Inc., New York 1949, S. 243

/13/ Glenn, E.E.; Slusser, M.L.: Factors Affecting Well Productivity II. Drilling Fluid Particle Invasion into Porous Media. Trans., AIME 1957, 210, S. 132-139

/14/ Tunn, W.: stimulationen bei tiefen Gasbohrungen. EEK-Z. 87 (1971), S. 202-208

/15/ Abrams, A.: Mud Design to Minimize Rock Impairment du to Particele Invasion. JPT, May 1977, S. 586-592

/16/ Ghofrani, R.; Mazeel Al Aboudi M.A. ; Sengupta, P. : Damage Caused by Clay-Based and clay-Free Inhibitive Pluids in Sandstone Formation. SPE, Feb. 1992, S. 439-445

/17/ Ghofrani, R.: Entwicklungsstand der Bohrspülungen für Übertiefe Bohrungen. Vortrag, Celle, 7.10. 1981

/18/ Ghofrani, R.: Formation Damage. NAM, Assen/Netherland, 1991

/19/ Allen, F.L.; Riley, S.M.; Strassner, J.E.: Initial Study of Temperature and Pressure Effects on Formation Damage by Completion Fluids. SPE 12488, 1984

/20/ Azari, M.; Leimkuhler, J.: Permeability Changes from Invasion of Sodium and Potassium-Based Completion Brines in Berea and Casper Sandstones. SPE 17149, 1988

/21/ Barna, B.A.; Patton, J.T.: Permeability Damage from Drilling Pluid Additives. SPE 3830, 1972

/22/ Davidson, D.H.: Invasion and Impairment of Formations by Particulates. SPE 8210, 1979

/23/ Dorsey, D.; Meyer, R.: Well Completion Technology Preventing Fluid Invasion Damage. Oil & Gas J. No. 50, 93, 81, 1983

-/24/ Earl , S. L. : Nahm, J.J.: Use of Chemical Salt Precipitation Inhibitors to Maintain Supersaturated Salt Muds for Drilling Salt Formation. SPE 10097, 1981

/25/ Gulbis, J.: Fracturing Fluid Chemistry. Drilling and Pumping Journal, Schlumberger Cambridge Research, July 1988, S. 4-21

/26/ Plank, J.P.; Gossen, F.A.: Visulation of Fluid-Loss Polymer in Drilling Mud Filter Cakes. SPE 19534, 1989

/27/ Hille, M.:

Vinylsulfonate/Vinylamide Copolymers in Drilling Fluids for Deep, High-Temperature Wells.
SPE 13558, 1985

/28/ Hille, M.: Hochtemperaturbeständige Wasserbasische Bohrspülungen.
EEK 97 (1981), 10, S. 371-374

/28/ Eaton, B.A.; Smithey, M.: Formation Damage from Workover and Completion Fluids.
SPE 3707, 1971

/29/ N.N: Principle of TLC Bridging Agents Multifrac Halliburton Services, Sales and Service Catalog Number 40

/30/ Robinson, B.M.; Holditch, S.A.; Whitehead, W.S.: Minimizing Damage to a Propped Fracture by Controlled Flowback Procedures.
SPE 15250, 1986, S. 487-500

/31/ Ershaghi, I.; Mehdi, A. : Modeling of Filter Cake Buildup under Dynamic-Static Conditions.
SPE 8902, 1980

/32/ Hashemi, R. ; Ershaghi, I.; Ammerer, N. : Proper Filtration Minimizes Formation Damage Oil & Gas J. No. 33, Aug. 1984 82, S.122-126

/33/ Hassen, B.R.: New Technique Estimates Drilling Filtrate Invasion.
SPE 8791, 1980

/34/ Hassen, B.R.: Solving Filtrate Invasion With Clay-Water Base Mud Systems.
World Oil No. 6, Nov. 1982 195, S.115-24

/35/ Holditch, S.A. et al. Effect of Mud Filtrate Invasion on Apparent Productivity on Drillstem Tests in Low Permeability Gas Formations.
SPE 9842, 1981

:

r

/36/ Kennedy, J.L. : Drilling Mud Particulates Can Cause Formation Damage.
Oil & Gas J. No. 31, Aug. 1971 69
S.62-64

/37/ Krueger, R.F.; Vogel, L.C. : Damage to Sandstone Cores by Particles from Drilling Fluids.
Drill & Prod. Prac., API 1954,
S. 158-171 [A]

/38/ Lampkin, R.E.: Skin Damage. Part 1: How Drilling Fluids Affect Formation Productivity.
Oil & Gas J. No. 48, Nov. 1966, S. 82-84

/39/ Lampkin, R.E.: Skin Damage. Part 2: How To Evaluate Permeability Damage in the Laboratory.
Oil & Gas. J. No. 49, Dec. 1966,
S. 119-125

/40/ Lauzon, R.V.: Colloid Sci Damage Problems.
Oil & Gas J. No. 31, July 1984, S. 175-179

/41/ N.N.: Skriptum zur Lagerstättentechnik.
ITE, TU, Clausthal, WS 1983/84

/42/ Schubert, H.: Untersuchungen von Trägerschädigung und -behandlung bei verschiedenen Poreninhalten am Beispiel des Bentheimer Sandsteins unter Anwendung neuer Meßmethodik Diplomarbeit, ITE, TU Clausthal

/43/ McKinney, L.K.: Formation Damage From Synthetic oil Mud Filtrates at Elevated Temperatures and Pressures.
SPE 17162, 1988

/44/ Morgenther, N.: Formation Damage Tests of High-Density Brine Completion Fluids.
SPE, Trans., AIME, 283, Nov. 1986

/45/ Peden, J.: Reducing Formation Damage by Better Filtration Control. Offshore Service Tech. No. 1, Jan. 1982, 15, S. 626-28,

/46/ Peen, J.M.; Avalo, M.R.; Arthur, K.G.: The Analysis of Dynamic Filtration and Permeability Impairment Characteristics of Inhibited Water-Based Muds. SPE 10655, 1982

-

/47 Hashemi, R.: / Perlmutter, B.A New Advancement for Determining the Degree of Filtration Necessary for Completion, Simulation, and Workover Fluids To Minimize Formation Damage. Proc., Oct. 1985, 2, S. 309-330

/48/ Prokop, C.L.: Radial Filtration of Drilling Mud PT, AIME, Vol. 195, 1952

/49/ Outmans, H.D.: Mechanics of Static and Dynamic Filtration in the Borehole. SPE, Texas, Jan. 1963

/50/ Poole, G.: Clear Water Brines Minimize Formation Damage Oil & Gas J., July 1981, 79, S. 151-161

/51/ Sharp, K.W.; Allen, B.T.: Pledger, T.M.: Filtration of Oilfield Brines - a Conceptual overview. SPE 10657, 1982

/52/ Wright, T.G.; Dorsey, D.: Formation Damage and Means of Prevention Using Workover and Completion Fluids. Proc. Lobbock, TX 1984, S. 72-75

/53/ Pence, S.A.: Evaluating Formation Damage in Low Permeability Sandstone. SPE 5638, 1975

/54/ Raibile, C.J.;
Gall, B.L.: Laboratory Formation Damage Studies of Western Tight Gas Sands.
SPE 13903, 1985

/55/ Krueger, R.F.: An Overview of Formation Damage and Well Productivity in Oilfield Operations
JPT, Trans., AIME, 281, Feb. 1986, S. 131-52

/56/ Amaefule, et al.: Advance in Formation Damage Assessment and Control Strategies.
CIM 88, Calgery (1985), 39

/57/ Bergosh, G.L.;
Enniss, D.O.: Mechanisms of Formation Damage in Matrix Permeability Geothermal Wells.
SPE 10135, 1981

/58/ Broaddus, G.: Well- and Formation-Damage Removal with Nonacid Fluids.
JPT, Trans., 285, Jun 1988, S. 685-697

/59/ Gulbis, J.: Dynamic Fluid Loss of Fracturing Fluids.
SPE 12154, 1983

,

/60/ ohen , H
.A . civan, P.: Simulation of Formation Damage in Petroleum Reservoirs.
SPE 19420, 1990

/61/ Monaghan, P.H., et al: Laboratory Studies of Formation Damage in Sands Containing clays.
Trans., AIME 216, 1959, S. 209-225

/62/ Rahman, S.: Untersuchung von Trägerschädigung durch Bohrspülungen. Dissertation, ITE, TU, Clausthal 1984

/63/ Beeson, M.C.; Wright, H.C.: Loss of Mud Sold to Formation Pores. JPT, August 1952, S. B-40-52

/64/ Marx, C.; Rahman, S.: Evaluation of Formation Damage caused by Drilling Fluids, Specifically in Pressure-Reduced Formations. JPT, Nov. 1987, S. 1449-1452

/65/ Hartmann, A.; Özerler, M.; Marx, C.: Untersuchung der Struktur von Filterkuchen aus Bohspülungen. Bericht, ITE, TU Clausthal 1985

/66/ Mahajan, N.C.; Barron, B.M.: Bridging Particle Size Distribution: A Key Faktor in the Desining of Non-Damaging Fluids. SPE 8792, 1980

/67/ Braunston, R.J.: Investigation of Well Damage Case History. SPE 10040, 1982

/68/ Wei, K.K.; Morrow, N.R.; Brower, K.R.: The Effect of Fluid, Confining Pressure, and Temperature on Absolute Permeabilities of Low-Permeability Sandstones. SPE 13093, 1984

69/ Renard, G.; Dupuy, J.G.: Influence of Formation Damage on the Flow Efficiency of Horizontal Wells. SPE 19414, 1990

/70/ Raible, C.: Formation Damage Studies of Low Permeability Sandstone. Report NIPER-13, ITT Research Inst., Bartlesville, 1986

/71/ Ghofrani, R.: Trägerschonende Spülunge
Vortrag, 7-21-Mai, Istanbul/Türkei
Scanned with CamScanner

/72/ Thomas, D.C.; Hsing, H.; Menzie, D.E.: Evaluation of Core Damage caused by oil-Based Drilling and Coring Fluids.
SPE 13097, 1984

/73/ Lecoutier, J.; Chauveteau, G.: Propagation of Polymer Slugs Through Porous Media.
SPE 13034, 1984

/74/ Thomas, D.C.: Thermal Stability of Starch and Carboxymethyl Cellulose Polymers Used in Drilling Fluids.
SPE 8463, 1979

/75/ Pye, S.D.; Smith, W.A.: Fluid Loss Additive Seriously Reduces Fracture Proppant Conductivity and Formation Permeability.
SPE 4680, 1973

/76/ Howard, G.C; Fast, C.R.: Hydraulic Fracturing.
SPE of AIME, Monograph Vol. 2, 1970

/77/ Montgomery, T.; Holditch, S.A.; Berthelot, J.M.: Effects of Fracture Fluid Invasion on Clean Behavior and Pressure Buildup Analysis. SPE 20643, 1990

/78/ Penny, G.S.: Nondamaging Fluid Loss Additive for use in

Hydraulic Fracturing of Gas Wells.
SPE 10659, 1982

/79/ Anderson, R.W.; Woodroof, R.A.: Synthetic Polymer Friction Reducers Can Cause Formation Damage.
SPE 6812, 1977

/80/ Volk, L.J. et al.: A Method for Evaluation of Formation Damage Due to Fracturing Fluids.
SPE 11638, 1983

/81/ Gall, B.L. et al.: Permeability Damage to Natural Fractures Caused by Fracturing Fluid Polymers.
SPE 17542, 1988

/82/ Sparlin, D.D.; Hagen, R.W.: Soluble Fluid-Loss Additives Can Reduce Well Productivities and Prevent Complete Gravel Placement.
SPEPE, Trans. AIME 285, Feb. 1988

/83/ Chatterji, J.; Borchardt, J.K.: Applications of Water-Soluble Polymers in Field.
JPT, 1981, 11, S. 2042-2056

/84/ Githens, C.J.; Burnham, J.W.: Chemically Modified Natural Guam for Use in Well stimulation.
SPE 5706, 1976

/85/ Stromberg Brown, D.; Curtice, R.J: Modeling the Effect of Time, Temperature, and Shear on the Hydration of Natural Guar Gels.
SPE 21857, 1991

/86/ Anderson, H.; Bratrud, T.P.; Delorey, J.R. A Logical Approach to Fracture Pluid Selection. Petroleum Society of CIM, Paper No. 32 (1981), 38

/87/ Sattler, A.R.; Hudson, P.J.; Raible, C.J.; Gall, B.L.: Laboratory Studies for the Design and Analysis of Hydraulic Fractured Stimulations in Lenticular, Tight Gas Reservoirs. SPE 15245, 1986

/88/ Puri, R.; King, G.E.; Palmer, I.D.: Damage to Coal Permeability During Hydraulic Fracturing. SPE 21813, 1991

/89/ Jacobs, I.C.; Thorne, M.A.: Asphaltene Precipition During Acid Stimulation Treatment. SPE 14823, 1986

/90/ Knobloch, T.S.; Farouq Ali, S.M.; Trevino Diaz, J.: The Role of Acid-Additive Mixtures on Asphaltene Precipitation. SPE 7627, 1978

.:
/91/ Roberts, L.D.; Sutton, G.D.: Paraffin Precipitation During Fracture Stimulation. SPE 4411, 1973

/92/ Penny, G.S.; Conway, M.W.; Lee, W.S.: The Control and Modeling of Fluid Leakoff During Hydraulic Fracturing. SPE 12486, 1984

/93/ Egbogah, E.O.: An Effective Mechanism for Fines Movement Control in Petroleum Reservoirs. CIM 35, Calgary (1984), 16

/94/ Gabriel, G.A.; Inamdar, G.R.: An Experimental Investigation of Fines Migration in Porous Media. SPE 12168, 1983

/95/ Gruesbeck, C Collins, R.E.: Entrainment and Deposition of Fine Particles in Porous Media. SPEJ, Dec. 1982, S. 847-856 [R]

.;

/96/ Khilar, K.C.; Folger, H.S.: The Existence of a critical Salt Concentration for Particle Release. J. Colloid Interface Sci. No. 1, Sept. 1984, 101, S. 214-224

/97/ Khilar, K.C.; Folger, H.S.: Sandstone Water Sensivity: Existence of a Critical Rate of Salinity Decrease for Particle Capture. Chem. Eng. Sci. No. 5, 1983, 38, S. 789-800

/98/ Kia, S.F.; Folger, H.S.; Reed, M.G.: Effect of pH on Colloidally Induced Fines Migration. J. Colloid Interface Sci. No. 1, July 1987, 118, S. 158-168

/99/ Stahlberg, W.: Untersuchung über die Fließeigenschaften und die Stabilisierung von Tonspülungen durch Schutzkolloide unter Bohrlochbedingungen. Dissertation, ITE, TU Clausthal, 1970

/100/ McDowell-Boyer, L.M.; Hunt, J.R.; Sitar, N.: Particle Transport Through Porous Media. Water Resources Res. No. 13, Dec. 1986, 22, S. 1901-1902

/101/ Sarkar, A.K.; An Experimental Investigation of Fines

Sharma, M.M.: Migration in Two-Phase Flow.
SPE 17437, 1988

/102/ Sharma, M.M.; Yortsos, Y.C.: Permeability Impairment Due to Fines Migration in Sandstones.
SPE 14819, 1986

/103/ Sharma, M.M.; Yortsos, Y.C.; Handy, L.L.: Release and Deposition of Clays in Sandstones.
SPE 13562, 1985

/104/ Sydansk, R.D.: Stabilizing Clays With Potassium Hydroxide
JPT, Aug. 1984, S. 1366-1374 [A]

/105/ Wojtanowicz, K.; Langlinais, P.; Krilov, Z.: Experimental Determination of Formation Damage Pore Blocking Mechanisms.
J. Energy Resource Tech. No. 1, March 1988, 110, S. 34-42

/106/ Wojtanowicz, K.; Krilov, Z.; Langlinais, P.: Study in the Effect of Pore Blocking Mechanisms on Formation Damage.
SPE 16233, 1987

/107/ Derby, J.A.; Simon: Formation Damage is Evaluated by New Instruments, New Methods.
oil & Gas J., 74 (Sep. 1976), 38, S. 209-214

/108/ civan, F.; Knapp, R.N.: Effect of Clay Swelling and Fines Migration Formation Permeability.
SPE 16235, 1987

/109/ Civan, F.; Knapp, R.N.; Ohen, H.A.: Automatic Estimation of Model Parameters for Swelling and Migration of Fine Particles in Porous Media.
AICHE Meeting, New Orleans, LA, March 1981

/110/ Civan, F.; Knapp, R.M.; Alteration of Permeability by Fine Particle Processes.

Ohen, H.A.: JPT, 3 (1989), 1, 2, S. 65-79

/111/ Hewitt, C.H: Analytical Techniques for Recognizing Water-Sensitive Reservoir Rocks. JPT (Aug. 1963), S. 813-818

/112/ Sharma, M.; Yortsos, Y.C.: Transport of Particulate Suspensions in Porous Media: Model Formulation. AICHE Journal, Vol. 33, No. 10, Oct. 1987, S. 1636-1643

/113/ Priishoim, S.; Nielsen, B.L.; Haslund, O.: Fines Migration, Blocking, and clay Swelling of Potential Geothermal Sandstone, Reservoirs, Denmark. SPE, 1987, S. 168-224

/114/ Rege, S.D.; Folger, H.S.: Network Model For Straining Dominated Particle Entrapment in Porous Media. Chemical Engineering Science, 42 (Sep. 1987), 7, S. 1553-1564

/115/ Gray, D.H.; Rex, R.H.: Formation damage in Sandstones Caused by Clay Dispersion and Migration. 14th National Conference on Clays & Minerals, Elmsford, NY 1966, S. 355-366

/116/ Vaidy, R.N.; Folger, H.S.: Fines Migration and Formation Damage: Influence of pH and Ion-Exchange. SPE 19418, 1990

/117/ Cusack, J. et al.: Diagnosis and Removal of Microbial/Fines Plugging in Water Injection Wells. SPE 16907, 1987

/118/ Lever, A.; Dawe, R.A.: Clay Migration and Entrapment in Synthetic Porous Media. Marine Pet. Geol. No. 2, May 1987, 4, S. 112-118

/119/ Holditch, S.A.: Factors Affecting Water Blocking and Gas

Flow from Hydraulically Fractured Gas Wells. JPT, Dez. 1979, 12, S. 1515-1524

/120/ Jones, J.T.; Neustadter, L.; Whittingham, K.: Water-in-Crude Oil Emulsion Stability and Emulsion Destabilization by Chemical Demulsifiers. JPT-Canada, April-June 1978

/121/ Walsh, M.P.; Lake, L.W.; Schechter, R.S.: A Description of Chemical Precipitation Mechanisms and their Role in Formation Damage During Stimulation by Hydrofluoric Acid. JPT, Sep. 1982, S. 2097-2112 [R]

/122/ Black, H.N.; Hower, W.F.: Advantageous Use of Potassium Chloride Water for Fracturing Water-Sensitive Formations. Paper API 851-39-F, 1970

/123/ Coon, R.M.; McDaniel, B.W.; Simon, D.E.: Evaluation of Fluid pH Effect on Low-Permeability Sandstones. SPE 6010, 1976

/124/ Coulter, G.R.; Hower, W.: The Effect of Fluid pH on Clays and Resulting Formation Permeability. Proc., Lubbock, TX., 1975, s. 115-118

/125/ Hall, B.E.: Workover Fluids: Part 5: Certain Chemicals React To Stabilize Clays and Fines in the Formation. World Oil 6 (Dec. 1986), 203, S. 49-50

/126/ Hower, W.F.: Influence of Formation Clays on the Flow of Aqueous Fluids. Proc., 1981, S. 117-127

/127/ Land, c.S.; Baptist, 0.C.: Effect of Hydration of Montmorillonite on the Permeability to Gas of Water-Sensitive Reservoir Rocks. JPT, AIME, 234 [A], Oct. 1965, S. 1213-1218

/128/ Lauzon, R.V.: Chemical Balance Stops Formation Damage. Oil & Gas J. 36 (Sept. 1982) 801, S. 24-26

/129/ Leone, J.A.; Scott, B.M.:	Characterization Damage During Water Clay-cContent Reservoir. SPERE, Trans., AIME, 285, Nov. 1988 , S. 1279-1286
/130/ Omer, A.E.:	Effect of Brine Composition and Clay Content on the Permeability of Sandstone Cores, SPE 16520, 1987
/131/ 11tis, M.i Di dier, G.; Lareal, P.:	Comparison of the Inhibitory Action of and Guanidine Hydrochloride Solution on Montmorillonite Swelling. JPT, Aug. 1982, S. 514-521
/ 132 /Mclaughlin , H . E . et al.:	. Aqueous Polymers for Treating Clay 1n oij And gas producing formations . SPE 6008 , 1976
/133/ Scheurman, R.F. Bergersen, B.M. :	Injection Water Salinity, Formation Pretreatment, and Well Operate SPE 18461, 1989
/134/ Hower, F.W. :	Influence of Clays on the Production of Hydrocarbons. SPE 4785, 1974
/135/ Baij al, s,K. # Houchin, L.R.i Bridges, K.L. :	A Particle Approach to Prevent Formation Damage by High-Density Brines During the Completion Process. SPE 21674, 1991
/136/ Mungan, N. :	Permeability Reduction Through Changes ir pH and Salinity. JPT, Trans., AIME 234, Dec. 1965, S. 1449-1153
/137/ Flock, D. L. ; Raleigh, J.T.:	A Study of Formation Plugging With Bacteria. JPT, Trans., AIME, 234 (R], Fel 1965, S. 201-206
/138/ Geesey, G.G. j Mittelman, M.W.; Lieu, V.T. :	Evaluation of Slime Producing Bacteria in Oil Field Core Flood Experiments. App. and Env. Micro., Feb. 1987, S. 278-283

/139/ Ghalambor, A. et al.: Remedial Methods for Bacterial Formation Damage by Application of Oxidizers. SPE 14821, 1986

/140/ Hart, R.T.; Fekete, T.; Flocke, D.L The Plugging Effect of Bacteria in Sandstone Systems. The Canadian Mining and Metallurgical

/141/ Kalish, P.J. et al.: The Effect of Bacteria on Sandstone Permeability. JPT, July 1964, S. 805-814

/142/ Lappan, R.E.; Folger, H.S.: The Effects of Bacterial Polysaccharide Production on Formation Damage. SPE 19418, 1990

/143/ Nord, W.; Lyklema : Protein Adsorption and Bacterial Adhesion to Solid Surfaces: A Colloid-Chemical Approach. Col. and Sci., 1989, 38, S. 1-13

/144/ Clementz, D.M. et al. Stimulation of Water Injection Wells in the Los Angeles Basin Using Sodium Hypochlorite and Mineral Acid. JPT, 1982, S. 2087-2096

/145/ Shaw, J.C. et al: Bacterial Fouling in a Model Core System. App. and Env. Micro., 1985, 49, S. 693-701

/146/ Updegraff, D.M.: The Effect of Microorganisms on the Permeability and Porosity of Petroleum Reservoir Rock. Penn Well Publishing Co., Tulsa, OK, 1983, S. 37-44

/147/ Faraz, A.M.: Untersuchungen zur Abtrennung von Bohrklein im Feinstkornbereich aus einer synthetischen Tonspülungen mit Hilfe der Druckfiltration. Dissertation, ITE, TU Clausthal, 1990

/148/ Rehmer, K.P.: Behandlungsflüssigkeiten auf der Basis von Salzausfällungssuspensionen für Arbeiten in Gasfördersonden Wirkungsmechanismen und chemisch-physikalische Eigenschaften Dissertation, Freiberg, 1989

/149/ Veatch, R.W.: Current Hydraulic Fracturing Treatment and Design Technology. SPE 10039, 1982

/150/ Diezel, H.J.: Zur Entwicklung der Frac-Technik seit 1960. EEK-Z. 106 (1990), 11, S. 434-438

/151/ Cleary, M.P.: The Engineering of Hydraulic Fractures State of the Art and Technology of the Fracture. JPT, 1988, 1, S. 13-21

/152/ White, J.L.; Daniel, E.F.: Key Faktors in MHF Design. JPT, 1981, 8, S. 1501-1512

/153/ Brinkmann, F.W.; Elwerath B.: Status Report on Fracturing of Deep and Low Permeable Formations in West Germany, SPE/DOE-Paper No. 9852, 1981, S. 253-268

/154/ Klose, G.; Krömer, B.: Frac-Planung und -Behandlung Söhlingen 2-4. EEK-Z., 99. Jg., (1983), S. 181-188

/155/ Krömer, B.J.: Untersuchung von fractechnischen Planungsgrößen. Mitteilung aus dem ITE, TU Clausthal, Mai 1984

/156/ Slusser, M.L.; Rieckmann, M.: Fracturing Low Permeable Gas Reservoir. EEK-Z. 92 (1976), 3, S. 70-76

/157/ Brinkmann, F.; Furberg, H.D.; Schöber, K.: Frac-Behandlungen in tiefen, geringpermeablen Gaslagerstätten-derzeitiger Stand und weitere Aussichten. EEK-Z. 96 (1980), 2, S. 37-44

/158/ Gardner, D.C.; Bikerts, J.V.: Rheological Characterization of Crosslinked and Delayed Crosslinked Fracturing Fluids Using a Closed-Loop Pipe Viscometer. SPE 12028, 1983

/159/ Holcomb, D.L.: Minimizing Reservoir Damage by Ensuring Proper Additive Performance in Stimulation

Fluids. SPE 14991, 1986

/160/ Brinkmann, F.; Krömer, E.; Reinicke, K.M.: Methoden zur Bestimmung fracspezifischer Parameter aus dem Förderverhalten von behandelten Gasbohrungen. EEK-Z. 98 (1980), 6, S. 236-245

/161/ Almond, S.W.: Factors Affecting Gelling Agent Residue Under Low Temperature. 10658, 1982

/162/ Almond, S.W.; Bland, W.E.: The Effect of Break Mechanism on Gelling Agent Residue and Flow Impairment in 20/40 Mesh Sand. SPE 12485, 1984

163/ Elbel, J. et al.: Increased Breaker Concentration in Fracturing Fluids Results in Improved Gas Well Performance. SPE 21716, 1991

/164/ Grundmann, S.R.; Lord, D.L.: Foam Stimulation JPT, 1983, 3, S. 597-602

/165/ Nguyen H.T.; LaFontaine, M.: Effect of Various Additives on the Properties of the Clarified XC Polymer System. SPE 19404, 1990

/166/ Harestad, K.; Saasen, A.; Sterri, N.: Fluid Mechanical Analysis for Control of Gravel Pack Jobs. SPE 21672, 1991

/167/ Houchin, L.R.; Dunlap, D.D.; Hutchinson, J.E.: Formation Damage During Gravel-Pack Completions. SPE 17166, 1988

/168/ Bertaux, J.: Treatment Fluid Selection for Sandstone Aciding Permeability Impairment in Potassic Mineral Sandstones. SPEPE, Feb. 1989, S. 41-48

/169/ Bryant, S.L.; Buller, D.C.: Formation Damage From Acid Treatments. SPE 17597, 1988

/170/ Burkill, G.C.C.; Lichaa, P.M.: Successful Matrix Acidizing of Sandstones Requires a Reliable Estimate of Wellbore Damage
SPE 5590, 1975

/171/ Crowe, C.W.: Precipitation of Bydrated Silica From Spent Hydrofluoric Acid-How Much of a Problem Is It?.
JPT, Trans., AIME, Nov. 1986, 281 [R], S. 1234-1240

/172/ Crowe, C.W.: Evalution of Agents for Preventing Precipitation of Ferric Hydroxide Fron Spent Treating Acid.
JPT, April 1985, S. 691-95 [A]

/173/ Crowe, C.W.: Prevention of Undesirable Precipitates From Acid Treating Fluid.
SPE 14090, 1986

/174/ Crowe, C.H.; Minor, S.S.: Acid Corrosion Inhibitor Adsorption and Its Effect on Matrix Stimulation Results.
SPE 10650, 1982

/175/ Hille, A.D.; Sepehrnoori, K.; Mu, P.Y.: Design of the HC1 Preflush in Sandstone Acidizing.
SPE 21720, 1991

/176/ Williams, B.B. et al.: Acidizing Fundamentals.
SPE of AIME, Vol. 6, 1979

/177/ Shaughnessy, M ; Kunge, K.R.: Understanding Sandstone Acidizing Leads Improved Field Practices.
&
JPT, July 1981, S. 1196-1202

/178/ Thomas, R.L.; Crowe, C.N.: Matrix Treatment Employs New Acid System for Stimulation and Control of Fines Migration in Sandstone Formation.
JPT, Aug. 1981, S. 1491-1500 [A]

/179/ Darly, H.C.H.; Harrison, G.; Hartfield, A.H.:	Acid-Soluble Drilling, Completion and Workover Fluids. Proc., Lubbock, Tx., 1974
/180/ Tuttle, R.N.; Barkman, J.H.:	New Nondamage and Acid-Degradable Drilling and Completion Fluids. JPT, Nov. 1974, S. 1221-1226
/181/ Grim, R.E.:	Applied Clay Mineralogy. McGraw-Hill Book Co., New York-Toronto-London, 1962
/182/ Rogers, W. F.:	Composition and Properties of Oil Well Drilling Fluids, Revised Edition, Gulf Publishing Company, Houston, Texas, 1935
/183/ Pöge, S.:	Untersuchungen von temperaturbedingten Veränderungen der Rheologie von Bohrspülungen. Dissertation, ITE, TU Clausthal, 1989
/184/ van Olphen, H.; Fripiat, J.J.:	Data Handbook for Clay Materials and other Non-Metallic Minerals, 1. Aufl., Wiley, New York-London-Syney-Toronto, 1979
/185/ Lagaly, G.:	Colloids. Ullmann's Encyclopedia of Industrial Chem., Vol. A 7, VCH, Weinheim, 1986
/186/ Hoffmann, U.; Endell, K.; Wilm:	Kristallstruktur und Quellung von Montmorillonit. Kristallographie 86 (1933), S. 340-348
/187/ Müller-Vonmoos, M. et al.:	Tonmineralogie und Bodenmechanik. Mitteilungen d. Instituts für Grundbau und Bodenmechanik, ETH Zürich, Nr. 133, 1988

/188 /Ruchwein, R.A.I Word , D . W — Mechanism of Clay Aggregation by Polyelectrolytes. Soil Science 23, 1952, S. 405-492

/189/ Packeter, A. — Interaction of Montmorillonite Clays With Polyelectrolyten. Soil Science 01, 1957, S. 335-343

/190/ Schott, H.: — Deflocculation of Swelling Clays by Nonionic and Anionic Detergents. Journal of Colloid Science 26, 1968, 5. 133-139

/191/ Schwarz, H.: — Experimentelle Untersuchungen zur stabilisierung von tonsuspensionen mit Natriumcarboxymethylcellulose. Dissertation, IT, TU Clausthal, May 1986

/192/ Becke, C.N.: Brunton, G.: — X-Ray and Infrared Data on Hectorite Guanidines and Montmorillonite-Guanidines Clay and Clay Minerals, Vol. 8, Oklahoma Oct. 1959

/193/ Vogt, K.; Köster, H.: — Zur Mineralogie, Kristallchemie and Geochemistry einiger Montmorillonite with Bentonite. Clay Mineral 13, 1978, S. 25-43

/194/ N.N.: — API-Norm RP 13B

/195/ Van Olphen, H.: — An Introduction to Clay Colloid Chemistery, 2. Ed., Wiley, New York-Londin-Sydney- Toronto, 1977

/196/ Weiss, A. et al.: — Anorg. Allgem. Chemie 284, 1956, S. 247-271

/197/ Granquist, W.T.: Pollack, S.S.: — A Study of the Synthethesis of Hectorite clay and Minerals. Volume 8, Proceedings of the 8th Nat. Conf. on Clay and Minerals, Oklahoma Oct. 1959, S. 151 ff

/198/ Müller, H.: — Kolloid-Beih. 16, 1928

/199/ Welder, G.: — Lehrbuch der physical chemistry. 2. Auf., VCH Verlagsgesellschaft, Weinheim 1985

/200/ Müller, H. et al.: — Offenlegungsschrift DE 3631764 Al, Offenlegungstag 03.88

/201/ Lagaly, G.1 Fahn, R.t	*Ton and tonminerale. Ullmanns Enzyklopädie der technischen Chemie, Band 23, 4. Aufl., Verlag Chemie, Weinheim, 1983, 8. 311-326*
/202/ Brandley, W.F. et al.:	z. Kristallographie Mineral., Petrogr. Abt. A 97, 1937, S. 216
/203/ Stahl	Fest-Flüssig-Seminar, Inst. for technology. Verfahrenstechnik, T.H. Karlsruhe, 1968
/204/ Stauff, J.:	Kolloidchenie. Springer Verlag, Berlin, 1960
/205/Ghofrani, R.:	Neuere Polymere in der Tiefbohrtechnik. Vortrag, Turkey, Mai1989
/206/ Williams, L.H.; Underdown, D.R.:	New Polymer offers Effective, Permanent Clay Stabilization Treatment. JPT, July 1981
/207/ Williams, L.H. David, R.; Underdown, R.:	New Polymer Offers Effective Permanent Clay Stabilization Treatment. SPE 8797, 1980
/208/ Shah, S.; Heinle, S.A. Glass J.E.:	Water-Soluble Polymer Adsorption From Saline Solutions. SPE 13561, 1985
/209/Moore, W.J.; Hummel, D.O.:	Physikalische Chemie. 2. Aufl., Walter de Gruyter, Berlin-New York, 1976
/210/ Hamker, H.C.:	Rec. trav. Chem. 55, 1936, S. 1015 and 56, 1937, S. 3
/211/ Overbeck, J.; Kruyt, H.R.:	Colloid, 1952

/212/v. Smoluchowski, M.: Physikalische Zeitschrift 17, 1916, S. 557-585

/213/v. Smoluchowski, M.: Zeitschrift für physikalische Chemie 92, 1918, S. 129 ff

/214/ Müller, H.: Kolloid-Zeitschrift 1, 1926

/215/ Lipatow, S.M.: Physikalische Chemie der Kolloid. 1. Aufl., Akademie-Verlag, Berlin, 1953

/216/ Kuhn, A.1 Kolloidchenischen Taschenbuch. Akademische Verlagsgesellschaft Geest Portig K.-G., Leipzig, 1960

/217/ Ghofrani, R.: Vorlesung, Tiefbohrtechnik III ITE, TU Clausthal

(218/ Kemper, K.: Geologischer Führer durch die Grafschaft Bentheim and die angrenzenden Gebiete. Nordhorn 1986, Heimatverein d. Grafschaft Benth. e.V., S. 172

/219/ Füchtbauer, H.: Zur Petrographie des Bentheimer Sandstein in Emsland. ERK-2. (1955), 8, S. 616-617

220/ Hartmann, A.: Untersuchungen a Filterkuchen von Bohrspülungen. Dissertation, ITE, TU Clausthal, 1987

/221/ Holub, R.W. ot al.: Scanning Electron Microscop Pictures of Reservoir Rocks Reveal Ways to Increase oil Production.

SPE 4787, 1974

/222/ Pittman, E.D. Thomas, J.B.: Some Application of Scanning Electron Microscopy to the Study of Reservoir Rock. JPT, Nov. 1979, S. 1375-1380

/223/ N.N: Haake is Viscosimeter. Betriebsanleitung, 1950

/224/ N.N.: Methods for the Determination of the Microstrukture of Porous Substancees. Carlo-Erba, Milano, Italy

/225/ Ghofrani, R., Alaboudi, M. M.,
& Sengupta, P. Damage caused by clay-based and clay-free inhibitive fluids in sandstone formations. Paper presented at the *SPE International Conference and Exhibition on Formation Damage Control,* SPE-MS. (1992).

/226/Al-Aboudi, M. M., Ghofrani, R., &
Sengupta, P. Damage caused by clay-based and clay-free inhibitive fluids in sandstone formations. *Journal of Canadian Petroleum Technology, 34*(01). (1995).

/227/Halim, M. C., Hamidi, H., &
Akisanya, A. R. Minimizing formation damage in drilling operations: A critical point for optimizing productivity in sandstone reservoirs intercalated with clay. *Energies, 15*(1), 162. (2021).

/228/Larestani, A., Mousavi, S. P.,
Hadavimoghaddam, F., &
Hemmati-Sarapardeh, A. Predicting formation damage of oil fields due to mineral scaling during water-flooding operations: Gradient boosting decision tree and cascade-forward back-propagation network. *Journal of*

Petroleum Science and Engineering, 208, 109315. (2022).

/229/Klungtvedt, K. R., & Saasen, A. The role of particle size distribution for fluid loss materials on formation of filter-cakes and avoiding formation damage. *Journal of Energy Resources Technology, 145*(4), 041702. (2023).

Biography of the Author

Dr.Eng. Dr. Management Muhammed Mazeel al-Aboudi has worked for many years in the international sector of the oil and gas industry in several countries. He studied at the Faculty of Mining and Geology-Department of Petroleum engineering of the University of Belgrade (Diploma in Engineering), the Escuela Superior de Minas in Madrid, Spain (Credit in Applied Geophysics Engineering) and the University of Clausthal-Zellerfeld in Germany. He was awarded a PhD in Petroleum Engineering by the Technical University of Clausthal-Zellerfied in Germany. He was awarded another PhD in Management - Marketing by University of Bradford in United Kingdome. He started working as, Petroleum Engineer, Reservoir Engineer, Drilling, and Production Engineering, focusing on issues like Reservoir Engineering, Petroleum Engineering, Fiscal Regimes, Cost Estimation, Petroleum Economics, Strategic Planning, Portfolio Management, and Energy Policies. Dr Mazeel has been employed in many European companies (RWE-Dea, Repsol, EON) as a Senior Petroleum – Reservoir Engineer; He has also worked as Director General of the Iraqi Drilling Company (IDC) and Director General for the Oil Products Distribution Company (OPDC). At the same time, he was the oil industry adviser to the Iraqi Prime Minister in 2005-2006. Director General of Reservoirs and Fields Development Directorate (RFDD) at the Ministry of Oil in Iraq. He is the Author of many papers and books on petroleum engineering, fiscal regimes, management and marketing, energy policies for the Oil and Gas Journal, Petroleum Economics Journal, Middle East Economy Survey (MEES) Journal, Society of Petroleum Engineers (SPE) Journal, the Canadian Institute of Mining, Metallurgy and Petroleum (CIM) Journal, the German Oil-Gas Coal Magazine (EEZ) Journal, Washington Institute and Diplomica Verlag in Germany.